Metrics for Architects, Designers, and Builders

Metrics for Architects, Designers, and Builders

Martin Van Buren

VNR VAN NOSTRAND REINHOLD COMPANY
NEW YORK CINCINNATI TORONTO LONDON MELBOURNE

Copyright © 1983 by Van Nostrand Reinhold Company Inc.

Library of Congress Catalog Card Number: 82-8406
ISBN: 0-442-28889-1

All rights reserved. No part of this work covered by the copyright hereon may be reproduced or used in any form or by any means—graphic, electronic, or mechanical, including photocopying, recording, taping, or information storage and retrieval systems—without permission of the publisher.

Manufactured in the United States of America

Published by Van Nostrand Reinhold Company Inc.
135 West 50th Street, New York, N.Y. 10020

Van Nostrand Reinhold Publishing
1410 Birchmount Road
Scarborough, Ontario M1P 2E7, Canada

Van Nostrand Reinhold
480 Latrobe Street
Melbourne, Victoria 3000, Australia

Van Nostrand Reinhold Company Limited
Molly Millars Lane
Wokingham, Berkshire, England

15 14 13 12 11 10 9 8 7 6 5 4 3 2 1

Library of Congress Cataloging in Publication Data

Van Buren, Martin, 1921-
  Metrics for architects, designers, and builders.

  Includes index.
  1. Engineering—Handbooks, manuals, etc. 2. Metric system. I. Title.
TA151.V26 1982   389'.16   82-8406
ISBN 0-442-28889-1   AACR2

Preface

The United States officially joined the international metric community with the ratification of the *Metric Conversion Act of 1975* (Pub. Law 94-168); and with the conversion the United States became a participant in the International System of Units known as the SI (from the French Système International d'Unités). However, the conversion to the metric system of measure was decreed on a voluntary, not compulsory, basis. No deadline was specified for its fulfillment. Producers of goods, as well as architects, planners, and interior designers were free to choose their own time in taking the metric plunge.

When complete conversion is to take place remains an unknown. Many manufacturers of building products and furniture have been reluctant to undertake the costly task of altering dimensions and sizes to metric measure without justifiable market demands. Producers of fabricated building materials such as doors and windows find an ample market from individuals replacing parts and remodeling existing structures. This conventional market will continue despite metric conversion. Furniture makers' sales are derived primarily from retail outlets, office supply houses, and interior decorators. It is true that the conversion of furniture sizes to accepted SI metric-modular rules would complement the total metric package, but until metrically oriented construction is sufficient to create a profitable demand, furniture manufacturers will probably remain firm in their established practices.

Conversion to metric sizes poses another problem to producers. They must carry double inventories, stockpile dual sizes of each product, and in turn expand their warehouse facilities. The demand for traditional sizes will not cease even when metric conversion becomes a total reality.

There is no doubt that conversion to the metric system will be achieved, but not as promptly or consistently as many expect. The conversion will be a piecemeal performance. Architects and interior designers must therefore be prepared for an era of dual systems of measure. Some products will be available in metric dimension, while others will retain the customary feet-and-inches measurement, the latter will of course have to be converted to metrics in order to correlate with metrically dimensioned plans.

The International Organization for Standardization (ISO) has set forth standards for dimensional and modular coordination in building design, which will apply to furniture products as well. But with no prescribed transitional time schedule in this country, the question of when the product makers will make their changes remains.

One of the main objectives of this book is to provide representative lists of current dimensions with their metric equivalents of the most common building-related materials and furniture. In cases in which manufacturers' measurements differ as many dimensions as possible are listed for a single product.

Sooner or later, of course, most of the present dimensions will be adjusted to SI metric standards, the exception being those which must satisfy anthropometric demands.

Other objectives of the book are that of providing guidelines on architectural metric scale, suggesting means for visualizing metric sizes, and indicating approved SI recommendations for the presentation and use of metric units and symbols.

In one sense, the work is intended as an aid for the transitional period; in another sense, it is a basic handbook of the metric system as it applies to the work of architects, planners, and interior designers. A concise manual is needed to acquaint those in the building design and interior planning field with the conversion to the metrics system of measure.

MARTIN VAN BUREN

Acknowledgments

It isn't easy compiling a treatise on a technical subject such as metrics for architects and related designers as well as illustrative drawings. In this endeavor there were no staffs of advisors nor academic committees contributing. Only helpful adherents to the project, who were knowledgeable and very helpful.

My grateful thanks to the following.

Keyes D. Metcalf, librarian emeritus of Harvard University, for his advice on the chapter on "Metrics for Libraries."

Hoyt Galvin, Charlotte, North Carolina, former director of libraries of Charlotte and Mecklenburg County and prominent library consultant, for his advice on preparation of the manuscript.

Charles Fesmire, Charlotte, North Carolina, authority on graphic reproduction methods and systems, for his advice on format and illustrations.

F. H. Schmitt, FARA, German born and educated architect now of Charlotte, North Carolina, for his assistance in compiling metric conversion tables.

Forrest Wilson, author and professor of architecture, College Park, Maryland, for his counseling, encouragement, and advice in preparation of the work.

And by no means least, my wife Janie, who encouraged me and prompted the continuation of the work.

Contents

Preface	v
Acknowledgments	vii
1. Metrics for Designers	1
2. The Metric Timetable	6
3. What Is A Meter?	14
4. Conversion	21
5. Scale	37
6. Rules and Protocol	42
7. Visualization	46
8. Man Is the Measure	50
9. Metrics and the Architectural Environment	58
10. Metrics and the Business Office	79
11. Metrics in Restaurants and Bars	95
12. Metrics and Libraries	99
13. Metrics in Residences	112
14. Auditoriums and Assembly	129
15. Metrics and the Handicapped	133
Selected Bibliography	139
Index	141

Metrics for Architects, Designers, and Builders

1
Metrics for Designers

THE WAY IT WAS

The English system of measure is the offspring of many cultures. It is a hodgepodge of standards harvested over the centuries, now a system considered too obsolete and equivocal for our increasingly sophisticated and precise technology. In addition, the English system does not lend itself to the burst of worldwide trade which has occurred during the past half century; most of the trading nations have already discarded the system, including England.

In the beginning of human history, rudimentary units of measure were adequate. Parts of the human body such as the hand, arm, foot, and finger, became units of measure. Seeds and stones became measures of weight and volume. Time was calculated by celestial periods, such as "so many moons."

Eventually these early units of measure became grouped into systems, the Babylonian, Phileterian, Egyptian, Grecian Olympic, Roman, and ultimately English system. Incongruities have existed with all of them.

One of the first recorded units of measure, the ancient "cubit," carries several interpretations, the most common being the Egyptian definition as the distance from the elbow to the fingertips (about 18 inches). Fractions of cubits were based on the span or width of a hand and the width of a finger. The Egyptian "foot" was first picked up by the Greeks and then the Romans, who divided it into 12 "unciae" (inches). Thereby evolved the Roman duodecimal base of 12, which became a standard. Julius Caesar, with centurions on his mind, proclaimed a mile to be a thousand paces, or 5,000 Roman feet, a measure which Queen Elizabeth arbitrarily changed—centuries later—to 5,280 feet, thereby creating the English statute mile.

Another incongruity: the word "yard" allegedly comes from the Saxon word "gird," which means the circumference of a person's waist. A yard therefore became the length of a sash or girdle. However, there are some who claim that the yard unit of measure was decreed by King Henry I as the distance from his nose to the tip of his thumb. How a yard also came to equal 3 Roman feet is a little vague.

In the present English system there are still discrepancies. For instance, there are troy, avoirdupois, and apothecary weights. There are inconsistencies in the length of a mile: in addition to the statute mile, there is a geographical, or nautical mile; a survey mile; and formerly there was a German geographical mile which equaled 4 nautical miles. In volume measure, there is a difference between the United States standard gallon and the British imperial gallon.

Beyond the incongruities that have been passed down, there are muddled complexities in the English system of measure, as the following list shows.

- 4 inches equals one hand
- 9 inches equals one span
- 12 inches equals 1 foot
- 36 inches, or 3 feet, equals 1 yard
- 5½ yards, or 16½ feet, equals 1 rod (or "pole" or "perch")
- 40 rods equals 1 furlong
- 1 fathom equals 6 feet, or 2 yards
- 5,280 feet, or 1,760 yards, equals 1 statute mile
- 6,080 feet equals 1 nautical mile
- 6,080.27 feet equals 1 U.S. Coast Guard survey mile
- 3 statute miles equals 1 league

In liquid measure (4 gills = 1 pint = 16 fluid ounces) there is also apothecary (1 fluid ounce = 8 drams = 480 minims = 2 tablespoons = 6 teaspoons = 1/128 gallon), not to speak of dry measure (2 pints = 1 quart, and 8 quarts = 16 pints = 1 peck, which is ¼ bushel). And so on.

And to complicate mathematical calculations, measurements smaller than 1 inch reduce to awkward-to-handle fractions, such as 1/8, 3/16, 7/64, or 25/32 (the thickness of 1-inch lumber, dressed).

Is it any wonder that school children climb the walls memorizing all these data? In fact, most adults cannot recite the few equivalents listed above; some (not surprisingly) cannot spot a hand from a span, or a gill from a dram. Therefore, is it any wonder that the simpler and more precise international metrics system came into being? As it did. In France.

The French Revolution had become a raging tempest, and Louis XVI, King of France, was doomed. In January, 1793, he was guillotined in Paris, a deed that was witnessed by the beautiful but conniving Queen Marie Antoinette who was herself to meet the head-chopper in a few months. Louis XVI was then 39 years old. He died believing that the only notable legacy of his crown would be the Revolution. He was wrong. From the reign of Louis XVI was born the system of *le mètre*.

The metric system of weights and measures.

Sporadic attempts to standardize a system of measure were made as far back as 1670, but Louis XVI brought the matter to a head. In 1790, with the Revolution already in full swing, Louis decreed that the French Academy of Sciences "deduce an invariable standard for all measures." The concept of the metric system was created. Since his majesty was able to count to 10, he found the system delightfully simple; he approved.

The basic unit of length was the *mètre*, or "meter" (from the Greek word *metron*, meaning "measure"). One meter was to equal one ten-millionth of the earth's quadrant, or the distance from the north pole to the equator along the meridian running near Dunkirk, France, through Barcelona, Spain.

During the ensuing turbulent years in France—including Napoleon's Hundred Days—there was little interest in such mundane matters as metric measure. It was not until 1840 that France made use of the metric system compulsory. In 1875 the first International Metric Convention was held in Sèvres, France, where the Treaty of the Meter was formulated. The first International Bureau of Weights and Measures was created.

Following the convention, 17 nations signed the Treaty of the Meter.

In 1889 a new definition was established for the length of the meter, called the International Prototype Meter. This specified a meter as "a graduated line standard of platinum-iridium." (In 1960 the meter was redefined as follows: "The length equal to 1 650 763.73 wavelengths in vacuum of the radiation corresponding to the transition between the levels 2 p_{10} and 5 d_5 of the krypton 86 atom.")

By 1900 a total of 35 nations had officially adopted the metric system—but not the United States.

Fruitless efforts had been made in the United States to arouse interest in conversion to the metric system of measure. One such effort did result, in 1866, in an Act of Congress which made it "lawful throughout the United States of America to employ the weights and measures of the metric system in all contracts, dealings, or court-proceedings." This legalized the system but otherwise gave no more than lip service to it. No mandate of conversion was issued, so the 1866 edict went virtually unnoticed.

More than 100 years later, in 1968, the Congress authorized a study of the metric system by the National Bureau of Standards. The study was completed in 1971, and a report entitled "A Metric America—A Decision Whose Time Has Come" was submitted to the Congress. It was recommended that the United States adopt the metric system. Accordingly, a bill was introduced in Congress in 1974 which would authorize conversion of the United States to the metric system. The bill was met with howls of dissent and was soundly defeated, allegedly because of an amendment which called for the government to subsidize the purchase of metric-dimensioned tools for all small businesses and all tool-using workers. The cost of such a subsidy, it was reported, would be calamitous.

By then the United States had become one of five countries (and the only industrialized nation) still clinging to the outmoded English system of measure. The other nonmetric nations were Burma, Liberia, Brunei, and Yemen.

A revised bill, known as the Metric Conversion Act (Public Law 94-168), was passed and signed by President Ford in December 1975. This legitimized conversion of the United States to the metric system, but the transition was decreed on a voluntary—not compulsory—basis. No time deadline was demanded to complete the conversion, although a program of metric education was initiated and a United States Metric Board was established. That word "voluntary" caused some skepticism on the part of proponents of the bill.

The greatest impact resulting from the conversion was that the United States became a participant in the International System of Units, or SI (from Le Système International d'Unités). The SI was established by the General Conference on Weights and Measures in 1960 after considerable debate and revision. Its purpose was to clarify and define metric units and to formulate standards for usage of metric (SI) units. Today the SI remains the international authority in this regard. Of particular importance to architects and related designers is the recommended practice for the use of metric (SI) units in building design and construction.

Many people—building designers as well as laypersons—decry such a drastic and absolute change in our system of weights and measures from the familiar English system despite the mutual acceptance of the metric system by the rest of the world. "Why?" they ask.

Because it is essential. Because it is imperative that a common language of measurements be defined which is understood by all nations. Scientists and the medical profession have long been oriented to metric measure not only because it is precise but also because it provides consistent international usage and furnishes less chance of misunderstanding in exchange of information. In the broad fields of commerce and manufacture, international trade virtually demands a common system of measure: nations belonging to the European Common Market refuse to accept goods not produced and labeled in metrics. Architectural building products and furniture made to feet-and-inches dimensions must be converted to metric equivalents (which in turn, requires awkward planning because of oddball metric results) to fit foreign metric-dimensioned buildings. The square-peg–round-hole syndrome.

Building designers must accept and acclimate themselves to metric (SI) usage simply because there is no other choice. To belabor a cliche, the world gets smaller.

2
The Metric Timetable

Known as the English system of weights and measures, it was the accepted standard in the United States until the Congress passed the Metric Conversion Act of 1975. Ironically, England began exploring a departure from its own system of measure and conversion to the metric system in 1960 and was actively engaged in the conversion process by 1966. Canada followed in 1970. This left the United States isolated from a metric world, still adhering to a system the English had already discarded.

When Congress enacted the Metric Conversion Act, it left a loophole: the conversion process was to be voluntary. To dissident objectors, and manufacturers reluctant to participate in the conversion, it was an invitation to renege or at least stall. An orderly transition thus became an annoying obstacle. When the National Institute of Building Sciences in 1980 sponsored a symposium on the subject in Chicago, the session ended in unresolved debate. There was still no agreement, especially in the critical building industry, about a coordinated time schedule. Proponents of the change groaned, "When?"

Total conversion to the SI metric system can best be described as "eventual." The transition will not take place overnight, even though it is well under way in many industries; nor will hard conversion of all products and goods be accomplished simultaneously. There will continue to be overlaps and gaps in the time schedule.

The American National Metric Council projects 1985 as the date when the construction industry will be totally "metricated." This is a noble hope, based on an assumption, and not an assurance. Building design may be converted to metric modules and architects may plan accordingly, but initially at an awkward disadvantage. Compromise between two systems of measure will be the name of the game. Dimensional coordination in accordance with SI metric standards will be done without benefit of many metrically dimensioned materials and products, since timely synchronization of hard conversion is apparently an illusory dream. Each interrelated trade and industry must face its own complex set of conversion problems and cost factors.

What are the realities involved in the conversion process? They concern two separate viewpoints—that of the building designer, whose endeavors are creative; and that of the producer of goods and products, whose concern is manufacturing. Sometimes there is a conflict of interest and understanding between the two. Possibly the contradiction of viewpoints adds salt to the wounds of disagreement about achieving a correlated metric conversion time schedule. Architects insist they could begin designing buildings in metrics tomorrow if metric-dimensioned products were readily available; whereas building product manufacturers counter that they would be happy to oblige, and go through the costly pangs of hard conversion, if they were assured of a market. It is an impasse. Each protagonist has goals and valid arguments, but the counteracting jousting is without purpose, because the accomplishment of metric conversion with its simplified system can help each achieve the goals and resolve the arguments. Once the conversion is completed, there will no longer be cause for debate between designer and producer. All that is required is a collaborative time alignment of the metric conversion process. Perhaps this is an oversimplified hope.

The justification of the metric system is more than the establishment of a common international system of measure. In part it has to do with the logical simplicity of the system. Because metrics is a decimal system, design and dimensional coordination as well as documentation and interpretation of drawings are simplified—in any language.

Modular standardization of building products and fabricated materials also simplifies estimating; planning site layout; and ordering, transporting, and distributing materials. Diversity of material types and sizes is also reduced, which makes materials handling and inventory control not only easier but less susceptible to error. This does not mean that every structure must be a clone of every other structure. Modular coordination within a building project may be a desirable objective, but not at the expense of making all buildings identical. Standardization of dimensions of windows or doors does not mean that style, color, material, texture, construction, and design must be all the same. Nor does it mean that all bricks must be the same or laid in identical course patterns. Such speculations have led to the objection by some architects that standardization and dimensional and modular coordination might restrict and limit creative freedom. On the contrary, it offers renewed challenges based on organized principles. Orientation to new precepts takes time.

In repair and replacement of duplicate materials, one frustration of maintenance personnel is the provoking statement, "No longer available," or "This size is discontinued." Often makeshift expediency, such as trimming and sizing on the site, is required. Presently most building products are dimensioned according to the manufacturer's production convenience and are offered as standard; a dozen manufacturers of a product may propose a dozen different dimensional standards, none of which correlate with others. This poses a problem to the building designer, who must choose one dimension over others in specifying a selected product. Recommendations of the International Organization for Standardization (ISO) provide a broad range of dimensional options, but all within the context of accepted metric modules. Once hard conversion is done and manufacturers fully absorb the costs, many argue that they will not be financially burdened. With respect to sales potential, metric conversion will open up foreign markets to even the smaller manufacturers.

Manufacturers survive on demand for their goods. Rarely does a producer gamble substantial development and production costs in an attempt to *create* a market. It is a safer investment to find a ready market and fulfill it, which is why manufacturers are wary of redimensioning (which often means redesigning) readily marketable products on the possibility of a new system of measure—and moreover one that is merely voluntarily imposed. To producers, the transition can be a costly gamble.

- Present markets for customarily dimensioned products are ample and will not disappear even when all new construction is metrically oriented. Existing buildings will still require replacement parts and materials. In many instances, remodeling of older structures will follow traditional feet-and-inches dimensional standards and duplication of customary materials.
- Conversion to metric modular standards will thus mean *addition*, not substitution, of a line of goods. Demand for *both* systems of measure must therefore be accommodated.
- Tooling will be necessary to produce both accepted standards.
- Dual stockpiling of parts and goods will be required. Such minor items as metric screw threads, pipe dimensions, etc., must be duplicated.
- Double inventories will be unavoidable.

- There will be a need for additional, usually expensive, warehouse space to house the dual stockpiles of goods.
- Conversion to SI metric standards will require recalculation of tolerances, stresses, and so forth, as well as thorough testing of newly dimensioned products.
- In some cases, a dimensional change may call for a change in raw material stock, e.g., sheet or rolled steel to a new tolerance may demand a different gauge (thickness). In order to produce both customary and metrically dimensioned products, raw material supplies may have to be increased.
- Most manufacturers employ competent sales engineers. Even though many have had technical or engineering backgrounds, they must be indoctrinated in the particular field they represent. This involves sales orientation courses or seminars. A sales representative for a particular firm must "know his product cold" in order to merit his job. A conversion to metric dimension and the consequent transition means that such sales persons must be able to interpret drawings in metrics and be knowledgeable about their products when conferring with architects. They, too, must learn a dual system of measure.

Training courses absorb time and money. One more overhead expense. Builders claim that they can teach construction workers to tackle metric measurements in a couple of weeks. Architects maintain that draftsmen can pick up SI metric conventions and usage within a week. All of which adds up to more overhead and expense.

Who is going to pay?

Some builders insist that construction costs resulting from conversion to the metric system will increase rather than decrease, at least during the initial period of acclimatization. There will be an interval when some building products will be available in metric dimensions and other products in the customary system. Fitting and coordinating these variant dimensions is time-consuming. Faced with this uncertainty, it is to be expected that building contractors will add an appropriate safety factor in their estimates and bids. Construction of a sizable building often takes 2 or 3 years, and with the indefinite nature of voluntary conversion, there is no assurance about which materials and products will be available in either system or when.

One of the theoretical advantages of the metric system is that of presizing all building materials and products to a coordinated modular system in multiples of the preferred SI base of 100 millimeters (mm).

Cutting and trimming on the job was to have been minimized or eliminated. This has not proven to be the case. In England, which has been following a charted metric conversion program since 1966, there is as much on-the-site cutting and trimming as there was under the old customary system of measure. This time-consuming process will eventually be eliminated, but it raises the familiar question of when.

Take one example. A standard dimension of drywall or plywood is 48 in × 96 in (4 × 8), which equals 1219 × 2438 mm. ISO "preferred" standards recommend 1200 × 2400 mm (47.2 in × 94.5 in). Hence, if SI metric standards are specified, on-site trimming is required. Spacing of studs is customarily adapted to the 48-in width, i.e., 16 in and 24 in on centers, which equals 406 mm and 610 mm, respectively. SI recommendations round these dimensions to 400 mm and 600 mm, or 15.7 in and 23.6 in. More on-site trimming. Since the total SI metric "package" is dimensionally coordinated, every metric dimension relates to other metric measurements and cannot be regarded as a separate metric conversion. All dimensions are coordinated within a totally integrated building system.

Metric conversion of furniture sizes and dimensions is a vital part of the building conversion process. Some furniture makers argue that "loose," or occasional, furniture has no bearing on metric building modules. On the contrary, interior planning and arrangement of furniture often dictate the sizes, proportions, and relationship of areas. An office or a conference space cannot be architecturally planned unless the furniture requirements are known. In some cases modular units such as case goods or files must be fitted to precise spaces. Unless these are dimensioned in accordance with accepted metric modules, solutions can be awkward. For instance, a standard desk size is 30 in × 60 in, which equals 762 × 1524 mm. Integrating a dozen or more desks into a typing-pool area dimensioned in metrics can be a frustrating effort. Rounding the 30 in × 60 in size to a hard, or actual, 700 × 1500 mm measure equals 27.6 in × 59 in—a workable size much easier to correlate with (metric) building modules.

One category of furniture which cannot be plausibly converted to SI metric modular standards is seating. The design and proportions are based on anthropometric data (that is, measurements of the human body) and are adapted to factors of human comfort. Although it is feasible to round off the surface areas of work counters or desks several millimeters without sacrificing usability and convenience, the depth, height, pitch, and contour of seating for various intended purposes must relate to human form and posture. An alteration of a few millimeters can be critical.

When a nation undertakes conversion to the metric system, it has customarily prepared a detailed breakdown time schedule, especially in the construction industry. The purpose of such a schedule is to coordinate the time allowances and proposed dates for completion of conversion of various elements of planning, preparation, and implementation of each sector of the building process, including product conversion, systems orientation, and conversion of construction procedures to metric dimensioning. Scheduled publication of metric data is often prescribed along with dates for conversion of surveying and mapping and liaison of labor and services. There are overlaps in such schedules, of course, and some of the projected conversion completion goals are staggered; but the objective is eventual dimensional coordination of the total metric conversion process—a detailed master timetable.

In England, the charted program schedule was prepared by the British Standards Institution (BSI). In the United States the coordinating schedule was formulated by the American National Metric Council (ANMC).

As noted previously, the ANMC conversion schedule projects 1985 as the date when the construction industry will be totally metricated. This date is admittedly arbitrary. The important factor is that the ultimate M day—the date when the metric transition is completed—is derived from elapsed time rather than a specific date.

In other fields involving consumer products, producers have progressed rapidly in the conversion process by labeling goods, by mass or capacity, in metric equivalents. In some cases this has meant no more than labeling customary products in metric terms; in other instances, such as the soft-drink industry, capacities have even been "hard-converted" to liters.

An unrelated variety of consumer goods does not, however, demand a common time schedule for conversion. The marketplace can offer an infinite assortment of goods labeled in either customary or metric capacity or mass without immediate disaccord between the two systems of measure. Such is not true with respect to the building industry, where a great variety of building materials, products, and furniture must interact for modular and dimensional coordination. In the building industry, a coordinated conversion time schedule is consequently

12 METRICS FOR ARCHITECTS, DESIGNERS, AND BUILDERS

AMERICAN NATIONAL METRIC COUNCIL
CONSTRUCTION INDUSTRIES COORDINATING COMMITTEE
CONSTRUCTION INDUSTRIES METRIC CONVERSION TIMETABLE

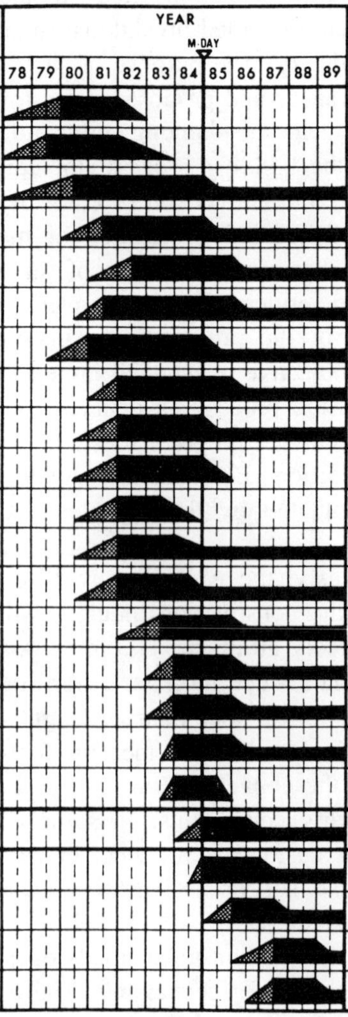

3.00 CONSTRUCTION INDUSTRIES COORDINATING COMMITTEE Action Item	YEAR (M-DAY) 78 79 80 81 82 83 84 85 86 87 88 89	Lead Sector	Other Sectors Involved
Publish Metric Practice Guide for Construction (and Revise)		3.02	3.01 3.03 3.04 3.05 3.06 3.07 3.08
Develop Necessary Basic Metric Standards		3.02	3.01 3.03 3.04 3.05 3.06 3.07 3.08
Determine Product Sizes in Metric Units		3.03	3.01 3.02 3.04 3.05 3.07
Decision to Build Metric Building		3.07	3.01 3.02 3.03 3.04 3.05 3.06
Award Design Contract		3.07	3.01
Develop and Distribute Metric Product Literature		3.03	3.09
Produce Metric Measuring Equipment		3.03	—
Begin Conceptual Design in Metric		3.01	3.02 3.03 3.04 3.07
Publish Model Codes in Metric		3.02	3.01 3.03 3.04 3.05 3.06 3.07 3.08
Develop Metric Estimating System		3.04	3.01 3.03 3.07
Enact Enabling Legislation		3.02	3.01 3.03 3.04 3.05 3.06 3.07 3.08
Conduct Land Transactions in Metric		3.06	3.02 3.07 3.08
Produce Property Surveys in Metric		3.08	3.01 3.02 3.04 3.06 3.07
Begin Design Development and Engineering in Metric		3.01	3.02 3.03 3.04 3.07
Estimate and Bid on Metric Documents		3.04	3.01 3.02 3.03 3.05
Award Metric Construction Contract		3.07	3.01 3.04
Accept Plans in Metric and Issue Building Permits		3.02	3.01 3.04 3.07
Obtain Labor Agreements in Metric		3.05	3.03 3.04 3.07
Begin Construction of Metric Buildings		3.04	3.01 3.03 3.05 3.07
Deliver Metric Products to Suppliers and Site		3.03	3.01 3.02 3.05
Complete Metric Construction		3.04	3.01 3.02 3.03 3.07
Occupy and Maintain Metric Buildings		3.07	3.01 3.03 3.04
Conduct Real Estate Transactions in Metric		3.06	3.01 3.02 3.04 3.05 3.07 3.08

Schedule Code: 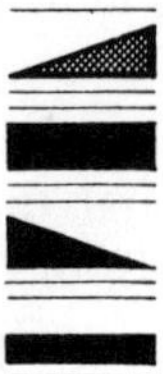
 Preliminary Metric Conversion Activity
More and more organizations become involved

Intensive Metric Conversion Activity
Most of the required changes are made

Decreasing Metric Conversion Activity
Any remaining changes are made

Normal Metric Activity
Metric measurement has become predominant

Sector Number Code:
3.01 Design
3.02 Codes and Standards
3.03 Construction Products
3.04 Contractors
3.05 Labor Liaison
3.06 Real Estate
3.07 Users/Building Clients
3.08 Surveying and Mapping
3.09 Information

Reprinted with permission from the ANMC Metric Reporter.

vital. Troublesome problems in architectural planning and construction will continue until all elements of the conversion schedule have been effected and conversion is no longer a part of the architect's task.

Builders are well aware of the immediate need for a coordinated metric conversion time schedule. Training of workers and the diversity of tools that they use will be greatly simplified.

The building industry is perhaps the only field in which many different trade and manufacturing endeavors are mutually dependent but must deal with metric conversion as uniformly as possible. For this reason, a realistic coordinated time schedule is very important. Although many segments of the building industry are actively undergoing—or planning—metric conversion, the task is far from being completed. Some of the pitfalls and problems of hard conversion on the part of manufacturers have been described here primarily to provide a realistic viewpoint and point out obstacles.

In some building product industries, conversion is relatively simple; in others, the switch involves complex changes in retooling, revision of specifications (and procurement) of raw materials, and sometimes complete redesign of a product. In one or two cases, there has been resentment that the industry is being "driven" into conversion at the expense of many to benefit a few.

The matter of voluntary-vs.-compulsory conversion remains a disturbing factor—to such an extent that it will frequently be mentioned in the text.

3
What Is a Meter?

There are more than 50 base units in the English system of measure. In the metric system there are 7 base units and 2 supplementary units, plus various derived units. Three of the base units are already in common use (time, electric current, and luminous intensity). Metric units constitute those in the following table.

QUANTITY	NAME	SYMBOL
Base Units		
Length	meter	m
Mass	kilogram	kg
Time	second	s
Electric current	ampere	A
Thermodynamic temperature	kelvin	K
Amount of substance	mole	mol
Luminous density	candela	cd
Supplementary Units		
Plane angle	radian	rad
Solid angle	steradian	sr

Since building design and its correlative manufactured materials and products deal primarily with measurement in length, the meter is the essential base unit to be considered inasmuch as it applies to architectural and interior planning. The terms of modular and dimensional coordination will follow SI recommended standards. Where no SI standards exist, as in the case of furniture products, suggested standards will be made for conversion.

The meter is not unfamiliar to most readers. Sports fans, especially of track and swimming, are familiar with 100 m, 400 m, 1000 m, and so on. Multiplied by a thousand, the meter becomes a kilometer (km), a measurement that is seen more and more on highway signs and is soon to be commonly accepted on survey maps. Divided by a thousand, the meter becomes the well-known millimeter (mm). In photography, 8, and 16, and 35 mm are standard film sizes. The 100s and

WHAT IS A METER? 15

120s cigarette lengths indicate millimeters. And a military artilleryman knows only too well the meaning of 90- , 105- , and 155-mm weapons.

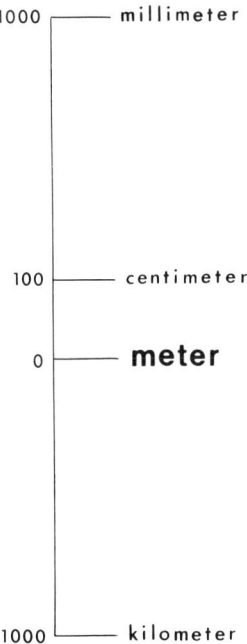

In the metric system, the meter is the base unit of linear measure. From this base unit, multiples and submultiples in powers of 1000 are preferred standards, with subdivisions in powers of 10 and 100. Metrics is thus a decimal system, like the United States monetary system: $0.01, $0.10, $1, $10, etc. The use of common fractions (numerator/denominator) becomes unnecessary.

Prefix symbols are standardized for all multiples and submultiples of the meter, as shown in the table.

PREFIX	MEANING	SYMBOL	VALUE, m
milli	millimeter	mm	0.001
centi	centimeter	cm	0.01
deci	decimeter	dm	0.1
	meter	m	1.0
deka	dekameter	dam	10
hecto	hectometer	hm	100
kilo	kilometer	km	1000

16 METRICS FOR ARCHITECTS, DESIGNERS, AND BUILDERS

This table, adequate for architectural and engineering practice, may be further simplifed as follows:

$$10 \text{ mm} = 1 \text{ cm}$$
$$10 \text{ cm} = 1 \text{ dm} = 100 \text{ mm}$$
$$10 \text{ dm} = 1 \text{ m} = 1000 \text{ mm}$$
$$10 \text{ m} = 1 \text{ dam}$$
$$10 \text{ dam} = 1 \text{ hm} = 100 \text{ m}$$
$$10 \text{ hm} = 1 \text{ km} = 1000 \text{ m}$$

In very precise scientific usage, it is often necessary to specify multiples larger than 1000 and subdivisions smaller than 0.001, for which the prefixes in the following table are accepted SI standards.

MULTIPLICATION FACTOR	PREFIX	SYMBOL
1 000 000 000 000 000 000 = 10^{18}	exa	E
1 000 000 000 000 000 = 10^{15}	peta	P
1 000 000 000 000 = 10^{12}	tera	T
1 000 000 000 = 10^{9}	giga	G
1 000 000 = 10^{6}	mega	M
1 000 = 10^{3}	kilo	k
100 = 10^{2}	hecto	h
10 = 10	deka	da
0.1 = 10^{-1}	deci	d
0.01 = 10^{-2}	centi	c
0.001 = 10^{-3}	milli	m
0.000 001 = 10^{-6}	micro	μ
0.000 000 001 = 10^{-9}	nano	n
0.000 000 000 001 = 10^{-12}	pico	p
0.000 000 000 000 001 = 10^{-15}	femto	f
0.000 000 000 000 000 001 = 10^{-18}	atto	a

There is no standard SI base unit for area, because these measurements are derived from the square of linear units, such as 1000 m² or 2250 mm², or are expressed in linear dimensions (150 × 200 m, or 500 × 2250 mm). In building practice, the width dimension precedes depth or height. Preferred terminology for square units is "square meter" or "square millimeter," rather than "meter squared" or "millimeter squared."

WHAT IS A METER? 17

Although the hectare is not a standard SI unit of area, its use is acceptable for surface measurement of land and water areas. The square meter (m²) is preferred for other area measure. The hectare (ha) equals 2.471 acres, or 10 000 m². Examples of area measure are large athletic fields (for football and baseball) which are noted in hectares, and materials such as carpet, which are measured in square meters.

The following are metric area equivalents:

$$100 \text{ mm}^2 = 1 \text{ cm}^2$$
$$100 \text{ cm}^2 = 1 \text{ dm}^2$$
$$100 \text{ dm}^2 = 1 \text{ m}^2$$
$$100 \text{ m}^2 = 1 \text{ dm}^2$$
$$100 \text{ dam}^2 = 1 \text{ hm}^2 \text{ (or ha)}$$
$$100 \text{ hm}^2 = 1 \text{ km}^2$$

Selection of appropriate prefixes in drawings and specifications should be in the most condensed form possible, preferably using numbers between 1 and 1000, but with as few decimal fractions as are feasible. For example, 5285 mm is more readable and less likely to be misinterpreted than is 5.285 m. In some calculations, however, it may be advantageous to use exponential notations rather than prefixes, such as

$$500 \text{ mm}^2 = 0.5 \times 10^{-3} \text{ m}^2$$

There is one submultiple of the meter—the centimeter—which is commonly seen and printed on such manufacturers' product sizes as package dimensions, clothing, furniture, and appliances, probably because the centimeter correlates so simply as a recognizable portion of an inch (2.54 cm = 1 in). That is, it is easy for an individual to calculate a centimeter as "a little less than half an inch." Moreoever, metric measuring rules and 1:1 architectural scales are divided into numbered centimeters and subdivided into 10 millimeters per centimeter (10 mm/cm). The numbered centimeters are easily read, totaling 100 per meter (100 cm = 1 m and 1000 mm = 1 m). It is thus logical for both building designers and laypersons to think in terms of centimeters. And for manufacturers to label their products in such an easily recognizable unit of measure.

Many marketed goods are presently labeled in both metric and English system measurements, a practice that is frowned on by metric advocates, because they see a tendency on the part of shoppers to read only the familiar English system labels without attempting to acquaint themselves with the accepted metric equivalents. Most producers maintain that they will in time drop dual labeling and tag their products only in metrics.

In building design and construction, however, the use of the centimeter prefix (cm, cm², cm³, etc.) in all forms as a unit of measure is not recommended and *should be avoided*. All dimensions and documents that indicate centimeters should be converted to millimeters or meters in building practice because:

- The centimeter is not consistent with the preferred SI use of multiples in ternary powers of 10.
- The difference between the magnitude of the millimeter and centimeter is only 10. The combined use of both units would lead to confusion and misinterpretation.
- To avoid misinterpretation and error in determining unit values, it is preferable to eliminate or minimize the extensive use of decimals or commas. Many European nations use commas as decimals. For accuracy, use of centimeters would require carrying digits to at least one decimal place (such as 2.5 cm or 3.8 cm), whereas the millimeter (mm) provides integers sufficiently accurate for building dimensions and related product measurements. Thus, decimal fractions can be eliminated.

Most furniture and many appliances correspond with building design and are affected by building dimensions. Conversion to metric sizes should therefore follow accepted SI architectural standards. Such products as modular office files, prefabricated cabinets and storage units, and appliances which must be fitted to prescribed spaces must therefore be adapted and converted to SI building standards. Most manufacturers who list their product sizes in metrics customarily do so in centimeters. Such dimensions should be converted to millimeters and indicated thus on drawings and specifications, with appropriate rounding to whole-millimeter measurements. In sections to follow, many of these products are listed and preconverted to millimeter dimensions.

Some common linear units with equivalent numerical values follow.

WHAT IS A METER? 19

Length

1 cable's length	120 fathoms
	720 feet (ft)
	219 m
1 cm	0.393 7 in
1 chain (surveyors)	66 ft
	20.1168 m
1 dm	3.937 in
1 dam	32.808 ft
	393.696 in
1 fathom	6 ft
	1.829 m
1 ft	0.3048 m
	304.8 mm
1 furlong	660 ft
	1/8 survey mile
	201.168 m
1 in	2.54 cm
	25.4 mm
1 km	0.621 mi
	3278.9 ft
1 league	3 survey mi
	4.828 km
1 link (surveyors)	0.66 ft
	0.201 168 m
1 m	39.37 in
	1.094 yards
1 micrometer (μm)	0.001 mm
	0.000 039 37 in
1 mil	0.001 in
	0.025 4 mm
1 statute mile	5280 ft
	1.609 km
1 nautical mile (nmi)	1,852 km
1 mm	0.039 37 in
	0.003 28 ft
1 nm	0.001 μm
	0.000 000 039 37 in
1 point (typography)	0.013 837 in
	0.351 mm

1 rod (pole or perch)	16.5 ft
	5.0292 m
1 yard (yd)	0.9144 m

Area

1 acre	43 560 ft²
	0.405 ha
1 ha	2.471 acres
	10 000 m²
1 cm²	0.155 in²
1 dm²	15.500 in²
1 ft²	0.092 903 m²
	92 903 mm²
1 in²	645.16 mm²
1 km²	247.105 acres
	0.386 mi²
1 m²	10.764 ft²
	1.196 yd²
1 mi²	258.999 ha
1 mm²	0.002 in²
1 rod²	25.293 m²
1 yd²	0.836 m²

This table of equivalents is intended for occasional ready reference. In the following chapter on Conversions, more detailed tables will be set forth.

There are two points to consider when converting to the metrics system which must be assimilated. One has to do with understanding of the system (the units of measure and correlation of metric units) as well as familiarity with SI preferred standards of usage and practice. In other words, we must have knowledge of the metric system itself.

The second has to do with interpreting and interrelating the numerical values of the customary English system of measure with metric equivalents, the often-awkward conversion factors. The problems of conversion will exist until total conversion is achieved.

4
Conversion

Many people long accustomed to the familiar English system of measure—including a number of architects and interior designers—grumble at the notion of adapting their mental and visual responses to the unfamiliar mechanics of metrics. The thought of terms, symbols, and numerical values accepted since childhood now to be superseded by a strange new system is annoying. When informed that metrics is basically a decimal system "as simple as dollars and cents," they scoff, "Simple you say? Don't give me that. I've got enough problems without getting involved in converting decimal fractions and all that . . . ."

The snag, of course, is not in the system itself but in the *conversion* from the English system to the metric system.

Many handbooks have been published for quick reference which explain the metric system in easy-to-understand and often general terms, such as,

- A millimeter is about the diameter of a paper-clip wire.
- A centimeter is about the width of a small fingernail.
- A meter is slightly less than bar-counter height.
- A kilometer is about 11 football fields, or a little more than half a mile.
- A gram is equal to the mass (weight) of a paper clip.
- A liter is a little more than a quart.
- A metric ton equals the approximate mass of a compact car.

And so on. Such metaphorical whimsies are helpful to the casual shopper or consumer but too imprecise for individuals involved in building, product design, and dimensioning. At the other extreme, some metric conversion tables are microcosmically explicit, intended for scientific and technological precision. But the span of millimeters in integers, without decimal fractions, provides adequate tolerance for most building design and correlative materials. The "diameter of a paper-clip wire" is a sufficiently accurate description for architectural dimensioning.

Until such time as the United States is totally and unequivocally adapted to the metric system in construction, building products, furniture, etc., architects, interior designers, and building contractors must face the aggravating task of converting from the English system to metrics.

Architects and interior designers fail to realize that if a building were designed from scratch in metric measure, if all the fabricated materials and furniture that went into the building were metrically dimensioned, there would be no conversion factors to contend with. Such is the case in countries where all materials, equipment, fittings, hardware, and furniture are of metric dimension. Immigrants to the United States with architectural background in the metric system often complain of the difficulty of adjusting to conversion from metrics to the more cumbersome English system. One German architect commented that it took several weeks to thoroughly assimilate the feet-and-inches enigma. "The decimal-oriented logic of the metric system is so rational," he explained. "Conversion to a duodecimal system is a frustrating effort."

By definition, conversion means a change in the form of a quantity without a change in value. This applies to conversion to the metric system of measure. In practical application to building design and construction, however, various factors must be considered, all of which have been previously noted, but are further defined here.

1. The terms soft and hard conversion of manufactured product sizes can be a booby trap to architectural and interior designers, because a misrepresentation can result in a dimensional error which might be critical. Soft conversion, which has to do with manufacturer's prerogative, is the culprit. The term means labeling the converted size of a product in conveniently rounded, or approximate, metric equivalent dimensions without changing the actual measurements. The product size remains unchanged. The object is to provide easily workable dimensions adaptable to metrically coordinated building standards.

> *Item:* A 48 in × 96 in (4′ × 8′) plywood or wallboard panel converts to 1219 × 2438 mm. For soft conversion, the same dimension may be labeled simply 1200 × 2400 mm.
>
> *Item:* A 30 in × 60 in standard office desk converted to metrics equals 762 × 1524 mm. This might be listed as 750 × 1500 mm.

Item: A square yard of carpet equals 0.836 m², which might be labeled simply 0.8 m².

In some cases, a minor inaccuracy may be within allowable building tolerances. In other instances, such as the thickness of wallboard or the details of exacting cabinetwork, precise conversion is vital. Also when a series of multiple units, such as rows or banks of lockers or filing cabinets, are specified, the additive difference between actual and rounded dimensional conversions can result in a substantial cumulative error. When a manufacturer lists inexact rounded numbers as metric conversion, the actual dimensions should be determined and the conversion values corrected.

The reasons for a manufacturer's choosing soft conversion are explained in Chapter 2.

Hard conversion (that is, when the true size of a product is changed to rational metric values) provides the most satisfactory and workable solution for building planners, contractors, and consumers. Thus, a measurement of 900 mm as listed will become an actual metric equivalent, or 35.4 in. A panel size of 1200 × 2400 mm will measure an actual 47.2 in × 94.5 in. The product size is changed to accord with its metric value. No corrections are necessary.

All manufacturers will eventually make appropriate modifications in *actual* sizes to whole-metric measurements (hard conversion) when the demand justifies the cost. This will simplify dimensioning and permit modular and dimensional coordination in all aspects of building design. Wherever possible, adjusted dimensions should be in powers of 10, in conformity with preferred metric values.

2. Rounding of numbers is intended to simplify dimensions within reasonable building tolerances. Under normal conditions, rounding millimeters to the nearest whole-digit number and meters to three decimal places, provide sufficiently accurate figures. Such rounding either eliminates or reduces unnecessary decimal fractions. In exacting situations which require more precise tolerances, a millimeter is seldom carried to more than two decimal places. Customarily, in the English system, a dimension is rarely more exact than 1/64 in, which converts to 0.396 875 mm. Only under the most demanding conditions would this metric conversion be carried to more than two decimal places: 0.40 mm in rounded numbers.

3. Conversion is essential to achieve a common system of measure.

24 METRICS FOR ARCHITECTS, DESIGNERS, AND BUILDERS

The selected system must be constant whether it is the English system, the metric system, or any other system. In building design, this includes all components relative to its planning including furniture and equipment. Buildings are frequently designed and constructed in foreign countries for which equipment and furniture are purchased in the United States. The structures are planned and dimensioned in metrics but the imported products are manufactured in feet and inches. Every item must be converted to metric measure in order to make accurate appraisals in planning and to correlate with the metric modules.

BASIC METRIC CONVERSIONS

1 in	= 25.4 mm	= 2.54 cm	= 0.025 m	
1 ft	= 304.8 mm	= 30.48 cm*	= 0.305 m	
1 yd	= 914.4 mm	= 91.44 cm*	= 0.914 m	
1 mi			= 1609.344 m	= 1.609 344 km
1 mm	= 0.039 in			
1 cm*	= 0.393 in			
1 m	= 39.370 in	= 3.281 ft	= 1.094 yd	
1 in²	= 645.16 mm²			
1 ft²	= 92 903.04 mm²		= 0.092 903 m²	
1 yd²			= 0.836 127 m²	
1 mi²				= 2.590 000 km²
1 mm²	= 0.001 550 in²			
1 cm²*	= 0.155 in²			
1 m²	= 1550.0 in²	= 10.764 ft²	= 1.195 99 yd²	

*Note: Inclusion of centimeters (cm) is for information only.

Never attempt to reconcile English system measurements with metric units; such a task is impossible. As a wise and witty engineer once said, speaking of machine tools, "An English nut and a metric screw will not easily mate."

Conversion tables are logical tools in the transition from the English system of measure to metrics. They will remain as vital references long after total conversion to metric measure is achieved—or for as long as traditional buildings exist which will require additional replacement materials and building products to coincide with the original specifications. For today's structure economists and investment planners calculate a financially feasible building life span of 20 to 30 years. Historical structures were designed on traditional principles—to last for generations. These traditional buildings will require traditional replacement components for maintenance or restoration.

Therefore, when metric conversion is complete, there will be a need for reverse conversion, that is, from metrics to English system customary dimensions. No doubt the next generation will look upon inches, feet, yards, quarts, pounds, and miles as alien terms which must be derived from metric equivalents. But conversion tables must and probably will work both ways.

To further simplify modular and dimensional coordination in building design and construction, the International Organization for Standardization (ISO) has formulated recommended preferred sizes for specific building materials and products. These are based on the SI basic module of 100 mm, and preferred multiples and submultiples, which are basis of overall building design coordination. For most products such as structural grids and many building components, the recommended size adjustments to preferred metric values present no transitional dilemmas; but for some products, there are problems.

People are growing taller: the height of the average adult increases by about 8 mm per decade. As a result, it is not uncommon today to see adult males well over 7 ft (84 in) tall romping the basketball courts. This equals 2134 mm in height, which is evidently not the maximum.

Makers of bedding have increased mattress lengths from the traditional 75 in (1905 mm) to 80 in (2032 mm) and even 84 in (2134 mm).

The average height of interior doors is customarily 80 in, which equals 2032 mm. Preferred metric dimensions for interior doors as recommended by ISO are:

2100 mm	which equals	82.7 in
2200 mm	which equals	86.6 in
2400 mm	which equals	94.5 in

Whether the metric door heights are a result of rounding of customary dimensions or that of the considered adoption of the above preferred measurements, it appears that such dimensional recommendations should be updated to take into account anthropometric sizes and proportions. Very tall persons confess to irritation when they must stoop to enter a room and a feeling of claustrophobia when low ceilings descend on them.

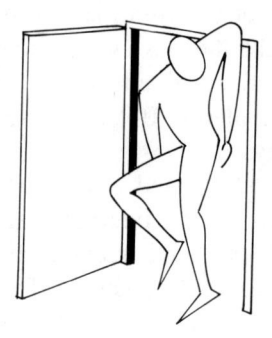

The following conversion tables on pages 27 through 36 represent the most common equivalent values in linear, area, and cubic measure.

INCHES AND FRACTIONS TO MILLIMETERS (1 in = 25.4 mm)

millimeters (mm)

Inches	0	1	2	3	4	5	6	7	8	9	10	11
0		25.40	50.80	76.20	101.60	127.00	152.40	177.80	203.20	228.60	254.00	279.40
1/16	1.59	26.99	52.39	77.79	103.19	128.59	153.99	179.39	204.79	230.19	255.59	280.99
1/8	3.18	28.58	53.98	79.38	104.78	130.18	155.58	180.98	206.33	231.78	257.18	282.58
3/16	4.76	30.16	55.56	80.96	106.36	131.76	157.16	182.56	207.96	233.36	258.76	284.16
1/4	6.35	31.75	57.15	82.55	107.95	133.35	158.75	184.15	209.55	234.95	260.35	285.75
5/16	7.94	33.34	58.74	84.14	109.54	134.94	160.34	185.74	211.14	236.54	261.94	287.34
3/8	9.53	34.93	60.33	85.73	111.13	136.53	161.93	187.33	212.73	238.13	263.53	288.93
7/16	11.11	36.51	61.91	87.31	112.71	138.11	163.51	188.91	214.31	239.71	265.11	290.51
1/2	12.70	38.10	63.50	88.90	114.30	139.70	165.10	190.50	215.90	241.30	266.70	292.10
9/16	14.29	39.69	65.09	90.49	115.89	141.29	166.69	192.09	217.49	242.89	268.29	293.69
5/8	15.88	41.28	66.68	92.08	117.48	142.88	168.28	193.68	219.08	244.48	269.88	295.28
11/16	17.46	42.86	68.26	93.66	119.06	144.46	169.86	195.26	220.66	246.06	271.46	296.86
3/4	19.05	44.45	69.85	95.25	120.65	146.05	171.45	196.85	222.25	247.65	273.05	298.45
13/16	20.64	46.04	71.44	96.84	122.24	147.64	173.04	198.44	223.84	249.24	274.64	300.04
7/8	22.23	47.63	73.03	98.43	123.83	149.23	174.63	200.03	225.43	250.83	276.23	301.63
15/16	23.81	49.21	74.61	100.01	125.41	150.81	176.21	201.61	227.01	252.41	277.81	303.21

FEET AND INCHES TO MILLIMETERS (1 ft = 304.8 mm; 1 in = 25.4 mm)

millimeters (mm)

Feet	0	1	2	3	4	5	6	7	8	9
0										
10	3 048	305	610	914	1 219	1 524	1 829	2 134	2 438	2 743
20	6 096	3 353	3 658	3 962	4 267	4 572	4 877	5 182	5 486	5 791
30	9 144	6 401	6 706	7 010	7 315	7 620	7 925	8 230	8 534	8 839
40	12 192	9 449	9 754	10 058	10 363	10 668	10 973	11 278	11 582	11 887
		12 497	12 802	13 106	13 411	13 716	14 021	14 326	14 630	14 935
50	15 240	15 545	15 850	16 154	16 459	16 764	17 069	17 374	17 678	17 983
60	18 288	18 593	18 898	19 202	19 507	19 812	20 117	20 422	20 726	21 031
70	21 336	21 641	21 946	22 250	22 555	22 860	23 165	23 470	23 774	24 079
80	24 384	24 689	24 994	25 298	25 603	25 908	26 213	26 518	26 882	27 127
90	27 432	27 737	28 042	28 346	28 651	28 956	29 261	29 566	29 870	30 175
100	30 480	30 785	31 090	31 394	31 699	32 004	32 309	32 614	32 918	33 223
110	33 528	33 833	34 138	34 442	34 747	35 052	35 357	35 662	35 966	36 271
120	36 576	36 881	37 186	37 490	37 795	38 100	38 405	38 710	39 014	39 319
130	39 624	39 929	40 234	40 538	40 843	41 148	41 453	41 758	42 062	42 367
140	42 672	42 977	43 282	43 586	43 891	44 196	44 501	44 806	45 110	45 415
150	45 720	45 726	46 330	46 634	46 939	47 244	47 549	47 854	48 158	48 463

FEET TO METERS (1 ft = 0.3048 m)

Feet	0	1	2	3	4	5	6	7	8	9
					meters (m)					
0	0.000	0.305	0.610	0.914	1.219	1.524	1.829	2.134	2.438	2.743
10	3.048	3.353	3.658	3.962	4.267	4.572	4.877	5.182	5.486	5.791
20	6.096	6.401	6.706	7.010	7.315	7.620	7.925	8.230	8.534	8.839
30	9.144	9.449	9.754	10.058	10.363	10.668	10.973	11.278	11.582	11.887
40	12.192	12.497	12.802	13.106	13.411	13.716	14.021	14.326	14.630	14.935
50	15.240	15.545	15.850	16.154	16.459	16.764	17.069	17.374	17.678	17.983
60	18.288	18.593	18.898	19.202	19.507	19.812	20.117	20.422	20.726	21.031
70	21.336	21.641	21.946	22.250	22.555	22.860	23.165	23.470	23.774	24.079
80	24.334	24.689	24.994	25.298	25.603	25.908	26.213	26.518	26.822	27.127
90	27.432	27.737	28.042	28.346	28.651	28.956	29.261	29.566	29.870	30.175
100	30.480	30.785	31.090	31.394	31.699	32.004	32.309	32.614	32.918	33.223
110	33.528	33.833	34.138	34.442	34.747	35.052	35.357	35.662	35.966	36.271
120	36.576	36.881	37.186	37.490	37.795	38.100	38.405	38.710	39.014	39.319
130	39.624	39.929	40.234	40.538	40.843	41.148	41.453	41.758	42.062	42.367
140	42.672	42.977	43.282	43.586	43.891	44.196	44.501	44.806	45.110	45.415
150	45.720	46.025	46.330	46.634	46.939	47.244	47.549	47.854	48.158	48.463
160	48.768	49.073	49.378	49.682	49.987	50.292	50.597	50.902	51.206	51.511
170	51.816	52.121	52.426	52.730	53.035	53.340	53.645	53.950	54.254	54.559
180	54.864	55.169	55.474	55.778	56.083	56.388	56.693	56.998	57.302	57.607
190	57.912	58.217	58.522	58.826	59.131	59.436	59.741	60.046	60.350	60.655
200	60.960	61.265	61.570	61.874	62.179	62.484	62.789	63.094	63.398	63.703

MILES TO KILOMETERS (1 mi = 1.609 km)

Miles	0	1	2	3	4	5	6	7	8	9
					kilometers (km)					
0		1.609	3.219	4.828	6.437	8.047	9.656	11.265	12.875	14.484
10	16.093	17.703	19.312	20.921	22.531	24.140	25.750	27.359	28.968	30.578
20	32.187	33.796	35.406	37.015	38.624	40.234	41.843	43.452	45.062	46.671
30	48.280	49.890	51.499	53.108	54.718	56.327	57.936	59.546	61.155	62.764
40	64.374	65.983	67.592	69.202	70.811	72.420	74.030	75.639	77.249	78.858
50	80.467	82.077	83.686	85.295	96.905	88.514	90.123	91.733	93.342	94.951
60	96.561	98.170	99.779	101.389	102.998	104.607	106.217	107.826	109.435	111.045
70	112.654	114.263	115.873	117.482	119.091	120.701	122.310	123.919	125.529	127.138
80	128.748	130.357	131.966	133.576	135.185	136.794	138.404	140.013	141.622	143.232
90	144.841	146.450	148.060	149.669	151.278	152.888	154.497	156.106	157.716	159.325
100	160.934	162.544	164.153	165.762	167.372	168.981	170.590	172.200	173.809	175.418
110	177.028	178.637	180.247	181.856	183.465	185.075	186.684	188.293	189.903	191.512
120	193.121	194.731	196.340	197.949	199.559	201.168	202.777	204.387	205.996	207.605
130	209.215	210.824	212.433	214.043	215.652	217.261	218.871	220.480	222.089	223.699
140	225.308	226.918	228.527	230.136	231.746	233.355	234.964	236.574	238.183	239.792
150	241.402	243.011	244.620	246.230	247.839	249.448	251.058	252.667	254.276	255.866
160	257.495	259.104	260.714	262.323	263.932	265.542	267.151	268.760	270.370	271.979
170	273.588	275.198	276.807	278.417	280.026	281.635	283.245	284.854	286.463	288.073
180	289.682	291.291	292.901	294.510	296.119	297.729	299.338	300.947	302.557	304.166
190	305.775	307.385	308.994	310.603	312.213	313.822	315.431	317.041	318.650	320.259
200	321.869									

SQUARE INCHES TO SQUARE MILLIMETERS
($1 \text{ in}^2 = 645.16 \text{ mm}^2$)

Square millimeters (mm^2)

Square Inches	0	1	2	3	4	5	6	7	8	9
0	6 452	645	1 290	1 935	2 581	3 226	3 781	4 516	5 161	5 806
10	6 452	7 097	7 742	8 387	9 032	9 677	10 323	10 968	11 613	12 258
20	12 903	13 548	14 194	14 839	15 484	16 129	16 774	17 419	18 064	18 710
30	19 355	20 000	20 645	21 290	21 935	22 581	23 226	23 871	24 516	25 161
40	25 806	26 452	27 097	27 742	28 387	29 032	29 677	30 323	30 968	31 613
50	32 258	32 903	33 548	34 193	34 839	35 484	36 129	36 774	37 419	38 064
60	38 710	39 355	40 000	40 645	41 290	41 935	42 581	43 226	43 871	44 516
70	45 161	45 806	46 452	47 097	47 742	48 387	49 032	49 677	50 322	50 968
80	51 613	52 258	52 903	53 548	54 193	54 839	55 484	56 129	56 774	57 419
90	58 064	58 710	59 355	60 000	60 645	61 290	61 935	62 581	63 226	63 871
100	64 516	65 161	65 806	66 451	67 097	67 742	68 387	69 032	69 677	70 322

AREA EQUIVALENTS (SURVEY MEASURE)

1 ft²	=	0.0929 m²	=	0.000 023 acres	=	0.000 000 036 mi²
1 m²	=	10.763 87 ft²	=	0.000 000 386 101 mi²	=	0.000 1 ha
1 acre	=	43 560 ft²	=	4046.873 m²	=	0.404 687 ha
1 ha	=	2.471 acre	=	0.003 861 mi²	=	10 000 m²
1 mi²	=	2 589 998 m²	=	640 acres	=	2.5900 km²

SQUARE FEET TO SQUARE METERS (1 ft^2 = 0.0929 m^2)

Square Feet	0	1	2	3	4	5	6	7	8	9
					Square meters (m^2)					
0	0.93	0.09	0.19	0.28	0.37	0.46	0.56	0.65	0.74	0.84
10	0.93	1.02	1.11	1.21	1.30	1.39	1.49	1.58	1.67	1.77
20	1.86	1.95	2.04	2.14	2.23	2.32	2.42	2.51	2.60	2.69
30	2.79	2.88	2.97	3.07	3.16	3.25	3.34	3.44	3.53	3.62
40	3.72	3.81	3.90	3.99	4.09	4.18	4.27	4.37	4.46	4.55
50	4.65	4.74	4.83	4.92	5.02	5.11	5.20	5.30	5.39	5.48
60	5.57	5.67	5.76	5.85	5.95	6.04	6.13	6.22	6.32	6.41
70	6.50	6.60	6.69	6.78	6.87	6.97	7.06	7.15	7.25	7.34
80	7.43	7.53	7.62	7.71	7.80	7.90	7.99	8.08	8.18	8.27
90	8.36	8.45	8.55	8.64	8.73	8.83	8.92	9.01	9.10	9.20
100	9.29	9.38	9.48	9.57	9.66	9.75	9.85	9.94	10.03	10.13
110	10.22	10.31	10.41	10.50	10.59	10.68	10.78	10.87	10.96	11.06
120	11.15	11.24	11.33	11.43	11.52	11.61	11.71	11.80	11.89	11.98
130	12.08	12.17	12.26	12.36	12.45	12.54	12.63	12.73	12.82	12.91
140	13.01	13.10	13.19	13.29	13.38	13.47	13.56	13.66	13.75	13.84
150	13.94	14.03	14.12	14.21	14.31	14.40	14.49	14.59	14.68	14.77
160	14.86	14.96	15.05	15.14	15.24	15.33	15.42	15.51	15.61	15.70
170	15.79	15.89	15.98	16.07	16.17	16.26	16.35	16.44	16.54	16.63
180	16.72	16.82	16.91	17.00	17.09	17.19	17.28	17.37	17.47	17.56
190	17.65	17.74	17.84	17.93	18.02	18.12	18.21	18.30	18.39	18.49
200	18.58	18.67	18.77	18.86	18.95	19.05	19.14	19.23	19.32	19.42
210	19.51	19.60	19.70	19.79	19.88	19.97	20.07	20.16	20.25	20.35
220	20.44	20.53	20.62	20.72	20.81	20.90	21.00	21.09	21.18	21.27
230	21.37	21.46	21.55	21.65	21.74	21.83	21.93	22.02	22.11	22.20
240	22.30	22.39	22.48	22.58	22.67	22.76	22.85	22.95	23.04	23.13

34 METRICS FOR ARCHITECTS, DESIGNERS, AND BUILDERS

SQUARE FEET TO SQUARE METERS (CONTINUED)

Square meters (m^2)

Square Feet	0	1	2	3	4	5	6	7	8	9
250	23.23	23.32	23.41	23.50	23.60	23.69	23.78	23.88	23.97	24.06
260	24.15	24.25	24.34	24.43	24.53	24.62	24.71	24.81	24.90	24.99
270	25.08	25.18	25.27	25.36	25.46	25.55	25.64	25.73	25.83	25.92
280	26.01	26.11	26.20	26.29	26.38	26.48	26.57	26.66	26.76	26.85
290	26.94	27.03	27.13	27.22	27.31	27.41	27.50	27.59	27.69	27.78
300	27.87	27.96	28.06	28.15	28.24	28.34	28.43	28.52	28.61	28.71
310	28.80	28.89	28.99	29.08	29.17	29.26	29.36	29.45	29.54	29.64
320	29.73	29.82	29.91	30.01	30.10	30.19	30.29	30.38	30.47	30.57
330	30.66	30.75	30.84	30.94	31.03	31.12	31.22	31.31	31.40	31.49
340	31.59	31.68	31.77	31.87	31.96	32.05	32.14	32.24	32.33	32.42
350	32.52	32.61	32.70	32.79	32.89	32.98	33.07	33.17	33.26	33.35
360	33.45	33.54	33.63	33.72	33.82	33.91	34.00	34.10	34.19	34.28
370	34.37	34.47	34.56	34.65	34.75	34.84	34.93	35.02	35.12	35.21
380	35.30	35.40	35.49	35.58	35.67	35.77	35.86	35.95	36.05	36.14
390	36.23	36.33	36.42	36.51	36.60	36.70	36.79	36.88	36.98	37.07
400	37.16	37.25	37.35	37.44	37.53	37.63	37.72	37.81	37.90	38.00
410	38.09	38.18	38.28	38.37	38.46	38.55	38.65	38.74	38.83	38.93
420	39.02	39.11	39.21	39.30	39.39	39.48	39.58	39.67	39.76	39.86
430	39.95	40.04	40.13	40.23	40.32	40.41	40.51	40.60	40.69	40.78
440	40.88	40.97	41.06	41.16	41.25	41.34	41.43	41.53	41.62	41.71
450	41.81	41.90	41.99	42.09	42.18	42.27	42.36	42.46	42.55	42.64
460	42.74	42.83	42.92	43.01	43.11	43.20	43.29	43.39	43.48	43.57
470	43.66	43.76	43.85	43.94	44.04	44.13	44.22	44.31	44.41	44.50
480	44.59	44.69	44.78	44.87	44.97	45.06	45.15	45.24	45.34	45.43
490	45.52	45.62	45.71	45.80	45.89	45.99	46.08	46.17	46.27	46.36
500	46.45									

CONVERSION

Acres to hectares (1 acre = 0.405 ha)

Acres	0	1	2	3	4	5	6	7	8	9
					hectares (ha)					
		0.40	0.81	1.21	1.62	2.02	2.43	2.83	3.24	3.64

Acres	0	10	20	30	40	50	60	70	80	90
					hectares (ha)					
0		4.05	8.09	12.14	16.19	20.23	24.28	28.33	32.37	36.42
100	40.47	44.52	48.56	52.61	56.66	60.70	64.75	68.80	72.84	76.89
200	80.94	84.98	89.03	93.08	97.12	101.17	105.22	109.27	113.31	117.36
300	121.41	125.45	129.50	133.55	137.59	141.64	145.69	149.73	153.78	157.83
400	161.87	165.92	169.97	174.01	178.06	182.11	186.16	190.20	194.25	198.30
500	202.34	206.39	210.44	214.48	218.53	222.58	226.62	230.67	234.72	238.76
600	242.81	246.86	250.91	254.95	259.00	263.05	267.09	271.14	275.19	279.23
700	283.28	287.33	291.37	295.42	299.47	303.51	307.56	311.61	315.65	319.70
800	323.75	327.80	331.84	335.89	339.94	343.98	348.03	352.08	356.12	360.17
900	364.22	368.26	372.31	376.36	380.40	384.45	388.50	392.55	396.59	400.64
1000	404.69									

36 METRICS FOR ARCHITECTS, DESIGNERS, AND BUILDERS

CUBIC FEET TO CUBIC METERS (1 ft^3 = 0.0283 m^3)

Cubic Feet	0	1	2	3	4	5	6	7	8	9
						Cubic meters (m^3)				
0		0.03	0.06	0.08	0.11	0.14	0.17	0.20	0.23	0.25
10	0.28	0.31	0.34	0.37	0.40	0.42	0.45	0.48	0.51	0.54
20	0.57	0.59	0.62	0.65	0.68	0.71	0.73	0.76	0.79	0.82
30	0.85	0.88	0.91	0.93	0.96	0.99	1.02	1.05	1.08	1.10
40	1.13	1.16	1.19	1.22	1.25	1.27	1.30	1.33	1.36	1.39
50	1.42	1.44	1.47	1.50	1.53	1.56	1.59	1.61	1.64	1.67
60	1.70	1.73	1.76	1.78	1.81	1.84	1.87	1.90	1.93	1.95
70	1.98	2.01	2.04	2.07	2.10	2.12	2.15	2.18	2.21	2.24
80	2.27	2.29	2.32	2.35	2.38	2.41	2.44	2.46	2.49	2.52
90	2.55	2.58	2.61	2.63	2.66	2.69	2.72	2.75	2.78	2.80
100	2.83									

5
Scale

Architects' measuring scales are familiar tools to anyone involved in building design. Such instruments are engraved with graduated divisions based on fractions of an inch: 1/32, 1/16, 1/8, 1/4, 1/2, etc. Each graduation on the selected scale represents 1 ft in actual measure. For identification, dimensions on drawings are labeled according to the selected scale, e.g., 1/8 in = 1'–0 in or 1/4 in = 1'–0 in That is in conformance with the English system of measure.

A metric measuring scale is simpler, because all ratio scales are based on a common decimal system. No multifaced instrument with varying fractional graduations for different scales is needed. All that is basically required is a graduated scale divided into centimeters and subdivided into 10 mm cm, which is actually nothing more than a metric rule. The numbered markings are centimeters (100 per meter). The 10 subdivision markings are millimeters. Thus, the number 1 on the scale equals 1 cm, which is equal to 10 mm; number 2 equals 2 cm, which is equal to 20 mm; and so on.

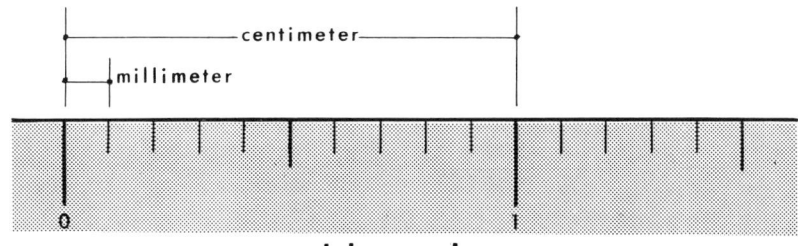

metric scale

The scale of a drawing is determined by ratio, which may be selected for convenience and legible interpretation. In the use of SI metric units, virtually any scale ratio can be selected which is readable and appropriate to the size of the drawing. One to one is full size, one to two is half size, and so on.

Metric scales are written with a colon designation (:) that means "to". Thus one to two is written as a scale of 1:2, or 1 to 100 is written as 1:100.

The ratio means that one unit on the scale equals the prescribed

38 METRICS FOR ARCHITECTS, DESIGNERS, AND BUILDERS

number of units shown in the unit ratio. For example, a scale of 1:100 means that:

- 1 mm on the scaled drawing equals an actual 100 mm.
- 10 mm on the scaled drawing equals 1000 mm, or 1 m.
- 1 m on a drawing equals an actual 100 m.

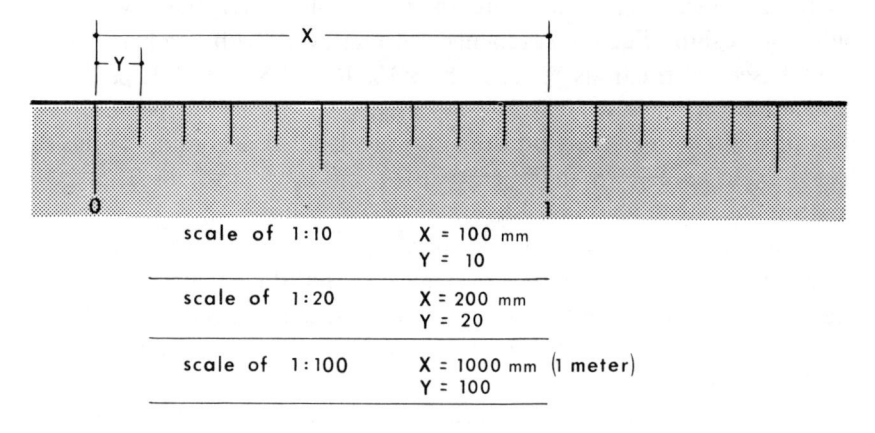

scale of 1:10	X = 100 mm Y = 10
scale of 1:20	X = 200 mm Y = 20
scale of 1:100	X = 1000 mm (1 meter) Y = 100

TYPICAL SCALES

Although any standard metric scale (or rule) as described above is adequate to determine any metric scale, most suppliers of drafting instruments offer architectural measuring scales calibrated to specific metric scale graduations. The common are:

        1:1   (Full size)
        1:5
        1:10
        1:20
        1:50
        1:100
        1:200
        1:500
        1:1500
        1:2000

Such architectural bar scales are of value to building designers who frequently use certain specific scale ratios. In addition, scales above 1:2000 are used by engineers; and for surveying and mapping, scales to 1:1 000 000 are employed.

SCALE 39

The following diagram and table indicates the calibration of available architectural bar scales and comparison with the standard metric rule, as well as the comparative application of metric scales to customary English system fractional scales.

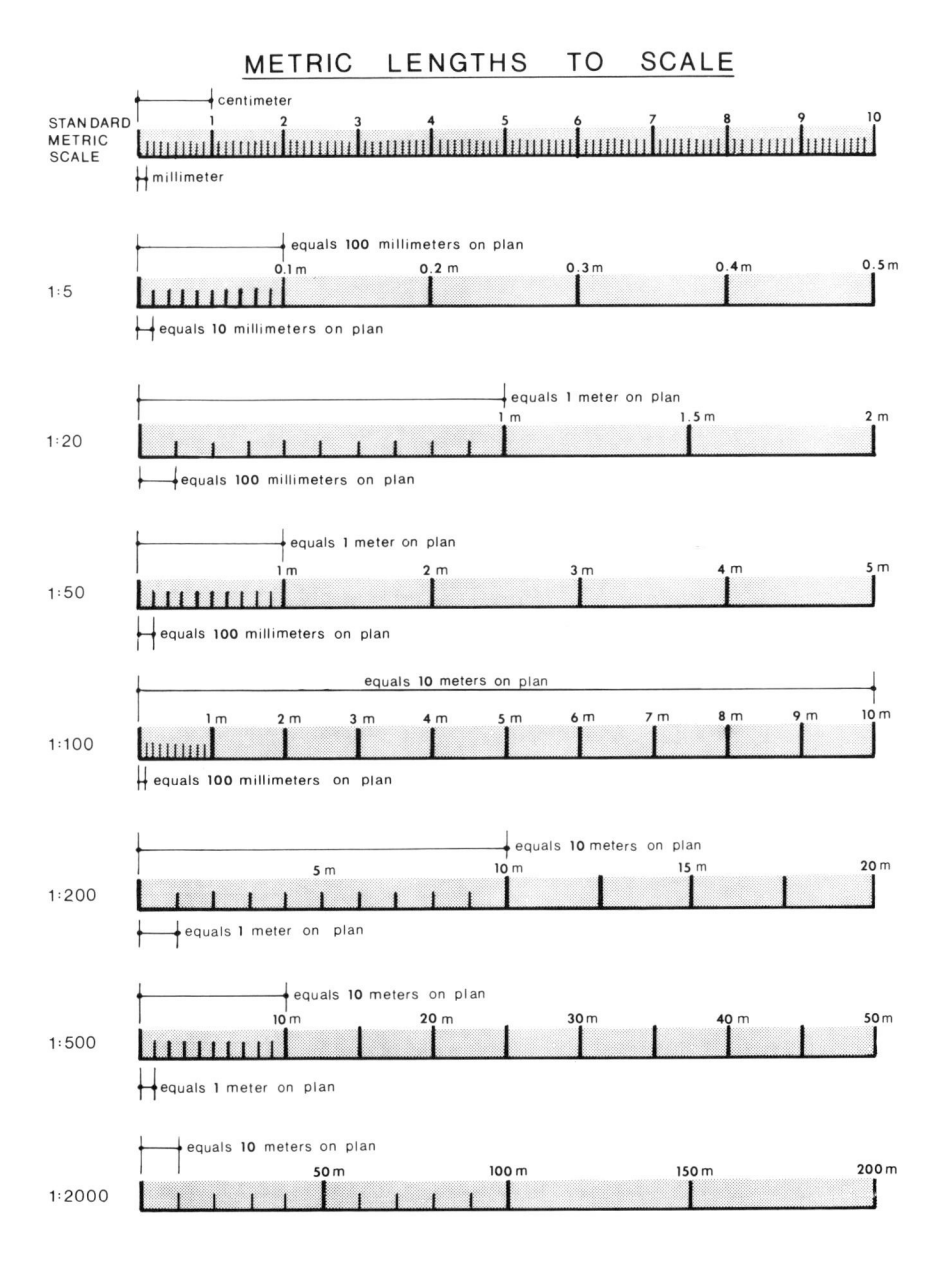

METRIC LENGTHS TO SCALE

STANDARD METRIC SCALE — centimeter / millimeter markings from 1 to 10

1:5 — equals 100 millimeters on plan; equals 10 millimeters on plan; 0.1 m, 0.2 m, 0.3 m, 0.4 m, 0.5 m

1:20 — equals 1 meter on plan; equals 100 millimeters on plan; 1 m, 1.5 m, 2 m

1:50 — equals 1 meter on plan; equals 100 millimeters on plan; 1 m, 2 m, 3 m, 4 m, 5 m

1:100 — equals 10 meters on plan; equals 100 millimeters on plan; 1 m, 2 m, 3 m, 4 m, 5 m, 6 m, 7 m, 8 m, 9 m, 10 m

1:200 — equals 10 meters on plan; equals 1 meter on plan; 5 m, 10 m, 15 m, 20 m

1:500 — equals 10 meters on plan; equals 1 meter on plan; 10 m, 20 m, 30 m, 40 m, 50 m

1:2000 — equals 10 meters on plan; 50 m, 100 m, 150 m, 200 m

Architectural Scale Ratios
English System Scales vs. Metric (SI) Scales

ENGLISH SYSTEM SCALE	APPLICABLE METRIC SCALE	PRECISE SCALE
1/32" = 1'–0"	1:200, or 1:500	1:384
1/16" = 1'–0"	1:200	1:192
1/8" = 1'–0"	1:100	1:96
3/16" = 1'–0"	1:50	1:64
1/4" = 1'''–0"	1:50	1:48
5/16" = 1'–0"	1:20, or 1:50	1:38
3/8" = 1'–0"	1:20, or 1:50	1:32
1/2" = 1'–0"	1:20	1:24
3/4" = 1'–0"	1:20	1:16
1" = 1'–0"	1:10	1:12
1½" = 1'–0"	1:10	1:8
3" = 1'–0"	1:5	1:4
Half size	1:2	1:2
Full size	1:1	1:1
Double size	2:1	2:1

The selection of the most appropriate scale for a drawing is a matter of common sense, logic, and experience. Most site plans that are comprised of an entire complex of buildings, such as an industrial center or several condominium structures, require drawings scaled to cover several hundred meters, suggesting a scale such as 1:2000. The most common scales for individual structures are 1:100 or 1:200. For detail drawings, such as those for furniture, scales of 1:5, 1:10, or 1:20 are normally used. Often a preliminary schematic which outlines nothing more than a basic plan is done at 1:400 or 1:500. Section details are commonly scaled at 1:20 or 1:50. Precise details, such as those for hardware assembly or cabinetwork joinery, may be half size (1:2) or full size (1:1). The object is to convey the desired information as directly and readably as possible.

It is important that preferred SI usage of metric units be followed, to eliminate misunderstanding and misinterpretation both endemically and internationally, which is one aim of the metrics system.

On each drawing or detail, stay with one unit of measure. Do not mix in one drawing meters with millimeters: this can lead to confusion in reading and interpreting dimensions. SI metric standards recommend that drawings scaled from 1:1 to 1:200 be dimensioned in millimeters; those from 1:200 to 1:2000, in meters, carried to no more than three decimal places. It is acceptable to note on a drawing: "All dimensions shown in meters (or millimeters)", in which case the unit symbol (mm or m) need not be shown after each dimension.

On a drawing in which all dimensions are shown in millimeters, five-digit numbers are acceptable.

When describing surface or sectional linear dimensions, indicate the width dimension first and then height or depth dimension.

In many European nations long accustomed to metric measure, architects and builders have accepted 100 mm as the basic dimensional module, with multiples of 2, 5, and 10 and powers of 10. They have found this procedure expedient and flexible, allowing ample variation in structural designing, dimensioning of bay modules, etc. This base module (100 mm) has been approved by the International Organization for Standardization (ISO) and will be adopted in architectural practice in the United States.

6
Rules and Protocol

Prescribed standards of usage and guidelines of practice in the use of metric units have been adopted by the International System of Units (SI). The purpose of coordinated presentation standards is to achieve conformity among nations in metric design and construction. Readability and understanding of building design drawings and specifications will be on common ground and will be clearly interpreted. The United States, now a participant in the SI, has accepted these standards and guidelines. It is recommended that they be followed in all building design drawings and specification documents.

Two examples may be given to illustrate the advantages of such standardization of usage of metric units. Both involve projects designed by foreign architects in foreign countries, for which the author (American) was retained as consultant and interior planner. Notations on drawings of one project were in Spanish; on the other, in Arabic; but the consistency in usage and presentation of metric dimensioning was clearly understood and interpreted on all plans and detail drawings. There was no need for explanation or further translation in correlating additional drawings.

Here are some pertinent selected rules and recommendations which are of special application to architectural and product design.

Metric Unit Symbols and Names

- Do not mix unit names and symbols within the context of a phrase or description. Use either "millimeters per square meter," or "mm per m^2"—*not* a mixture of name and symbol (such as "miilimeter per m^2").
- Never capitalize unit symbols. Use lowercase (m, mm, or km).
- Never use a period after a unit symbol except at the end of a sentence (m, mm, km).
- Use no space between letters in symbols containing more than one letter (mm, cm, km).

- Never change symbols in the plural (1 mm, 100 mm).
- Leave a space between digits and their related symbol (10 mm, 150 m, *not* 10mm, 150m).
- Show a zero before the decimal point for all numbers smaller than 1.0 (0.123, 0.005).
- Never use abbreviations for "square" (sq.) or "cubic" (cu.) (100 mm^2, 150 m^3; *not* 100 sq mm, 150 cu m).
- When written out in full, start all unit names with a lowercase letter except at the beginning of a sentence. (5 millimeters, 50 meters; *not* 5 Millimeters, 50 Meters).
- When the numerical value of a unit is written out in full, write the unit also out in full (fifty meters, ten millimeters; *not* fifty m, ten mm). However, it is proper, if desired, to follow number digits with the unit written out in full (50 millimeters, 10 meters).
- Leave a space on each side of signs of operation for multiplication, division, subtraction, and addition (5 m × 6 m = 30 m, 25 mm + 10 mm = 35 mm).
- Never hyphenate a unit name, even at the end of a line. (milli meter or kilometer; *not* milli-meter or kilo-meter)
- In numbers of more than four digits on either side of the decimal, arrange digits in groups of three from the decimal mark with a space between each group. Do not use commas (123 456.789 05, 0.987 650; *not* 123,456.789,05).
- For numbers with no more than four digits before or after the decimal use no spacing (1234, 0.5678).
- In all columned tabulations with five or more digits before and/or after the decimal, arrange digits in groups of three. 234 567 + 0.8765 + 25.6768 should align as follows

$$\begin{array}{r} 234\ 567 \\ 0.876\ 5 \\ \underline{25.676\ 8} \\ 234\ 593.553\ 3 \end{array}$$

- Avoid common fractions (numerator/denominator); use decimal notations for fractional numbers (3.25; *not* 3-1/4).

44 METRICS FOR ARCHITECTS, DESIGNERS, AND BUILDERS

- For area measure use preferred units—the square millimeter (mm²), the square meter (m²), and square kilometer (km²). The hectare (ha) is used for area measure of land and water only.
- For area and volume measure use the modifier before the unit name (square meter, cubic millimeter; *not* meter squared, millimeter cubed).
- In writing specifications, select the simplest appropriate prefix (150 m; *not* 150 000 mm).

Dimensioning Drawings

- Show the decimal mark in typing or printing at the level of the base line. When dimensions or numerals are handwritten, the decimal mark is slightly above the base line (25.8 mm, 25·8 mm)
- Use meters on all drawings scaled between 1:200 and 1:2000. Use millimeters on drawings scaled from 1:1 to 1:200.
- Avoid all use of centimeters including square and cubic centimeters. Where centimeters (cm) are shown on data or documents, convert to millimeters (mm) or meters (m).
- On each drawing or detail, show only one unit of measure throughout. Do not mix units of measure on a single detail drawing, e.g., millimeters and meters.
- Note that by labeling a drawing, "All dimensions shown in meters (or millimeters)," you can omit the unit symbol on that drawing (simply 6859, or 25.687).
- Note that whole numbers with no decimal always indicate millimeters (900, 1200, 4560).
- Note that decimalized numbers, taken to three decimal places, always indicate meters (0.123, 23.875, 0.005).
- Show area measure in square meters as m², or in square millimeters as mm². Square kilometers (km²) may be used for large areas. Never use square centimeters (cm²).

There are additional standards and recommendations for the use of SI metric units. However, the above are the most applicable to building design practice and related interior design.

RULES AND PROTOCOL 45

METRIC DIMENSIONING

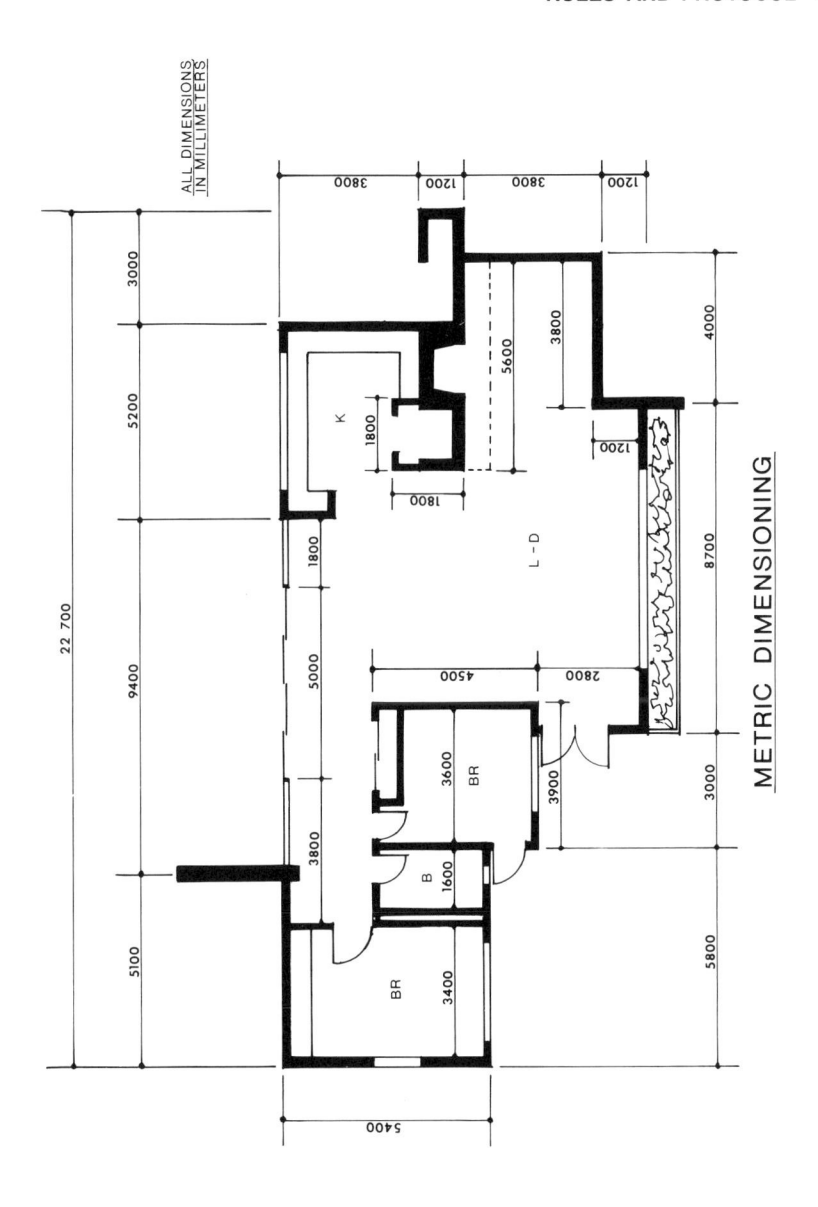

ALL DIMENSIONS IN MILLIMETERS

7
Visualization

The ability to envision spaces and objects in relative size and proportion is achieved by training. Seldom if ever is the capacity for creative visualization an inherent trait, although it is argued that occasionally certain genius mentalities intuitively possess the talent of creative vision. There may be varying aptitudes in learing to "picture in the mind," but visualization is a skill and any skill must be acquired even if it is self-taught.

In order to envision architectural elements and related objects in true perspective, the mind must be conditioned to accept some standard of measure which will serve as a basis for mental comparison and spatial judgment. Without a defined means of measurement, dimensional relations become discordant abstractions. Mental images have no actual common base.

In the United States, designers have been conditioned since childhood to the English system of measure. We have learned to dimension by this system, scale our drawings by this system, and think creatively by this system. The English system has been our standard of measure until now.

For a professional designer thoroughly indoctrinated in one system of measure, it is no easy task to shift gears and undergo reorientation to a completely different system of measure, i.e. units of metric measure. This is particularly true of the subconscious process of envisioning in a habit-formed standard of measure.

Understanding conversion tables and recommended SI standards is of little value to designers of buildings and the furnishings that go into buildings unless sizes and dimensions can be visualized in metric terms, which means seeing with the mind's eye spaces, sizes, proportions, and dimensions and their true relations in millimeters and meters, not merely envisioning in the habitual feet-and-inches system and mentally converting. In order to discuss a project in SI terms, a designer must also think in those terms. It would be an awkward (and perhaps embarrassing) gesture for a professional designer to have to produce a metric rule to compare measurements when discussing architectural dimensions in metric measure. Professional designers are supposed to be knowledgeable in such matters.

The necessity for designers to adjust to thinking in terms of metrics is more close at hand than many believe. Most schools now teach the metric system to students, who are already learning to "think metric"; and manufacturers are stepping up their efforts to convert products to metric measure. Architects and other designers in the building field are viewed as experts in their profession. Learning to think as well as plan in metric measure is already vital to their practice.

The facility of visualizing in a selected system of measure is learned from experience until the skill becomes second nature and the trained mind responds automatically. Define a room as 20 by 30 ft (20'-0" × 30'-0") and the presently trained mind will immediately fix a mental image of size and proportion. Or specify a four-drawer vertical file and its dimensions are clearly seen in feet and inches.

Such examples seem fundamental, until one is asked to envision that room and the vertical file in meters and millimeters. One must *think* of the room as a rounded 6 × 9 m (precise conversion is 6.096 × 9.144 m), and the prescribed file as 450 × 720 × 1330 mm (precise conversion is 457 × 724 × 1334 mm). Enter a room and *think* of that space as $XX \times YY$ meters. Think of a 30 in × 60 in desk as a rounded 750 × 1500 mm (exact conversion is 762 × 1524 mm).

As orientation exercises, the mental familiarization with metric measure can be aided by measuring spaces and objects with a metric rule or tape and orienting these dimensions mentally. Forget feet and inches and visualize the measurements *in metric terms*. Attempt later to judge metric dimensions by eye and then check the accuracy of your estimates by rule measurement. Enter an unfamiliar room—say an office—and look at its plan in terms of millimeters and meters. Estimate its overall size; ascertain how many millimeters the door is from the adjacent partition; guess the size of the desk and how many millimeters it is from the nearest wall. The size of window openings; the height of the sill; the metric dimensions of seating, tables, lamps, and wall ornaments; and so forth.

Mentally gauge the facades of buildings, their width and height, in terms of meters. Figure how many meters or millimeters a modular bay is. Visualize the metric dimensions of an athletic court or field.

Mental orientation is initially often aided by association with known reference dimensions. For example, a tall man (6 ft 6 in) is approx-

48 METRICS FOR ARCHITECTS, DESIGNERS, AND BUILDERS

imately 2 m, or 2000 mm, in height. The width of an average hand is about 100 mm. For larger-scale reference, a football field measures 91.440 × 48.768 m. Official competitive swimming pool length is 50 m. And so on.

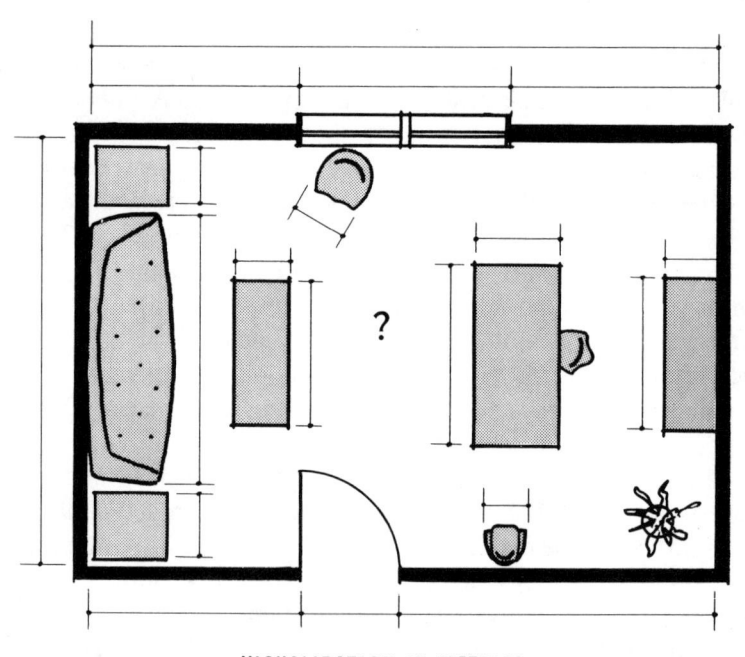

VISUALIZATION IN METRICS

Avoid estimating in feet and inches and then attempting to convert mathematically in your mind. *See* the dimensions in metric units. Avoid thinking of a room dimension of 3 m as a little more than 3 yd, or "about 10 ft". That kind of mental gymnastics is self-defeating.

Learn to *think* metric.

The discussion above undoubtedly seems primitively basic, even juvenile. But the reorientation of visual perception must start somewhere. In this approaching era of metric measure, it is the professional designer and architect who is expected to be the knowledgeable authority. It is a professional sin to disillusion a client's trust.

We see as we are taught to see in certain prescribed terms. The conceptual image of dimensional and spatial relations is adapted to a familiar unit of measure, whatever it might be. Visualization thus results from a series of optical experiences to which we become adjusted, so that mental images respond accordingly. It is a form of conditioned reflex, reacting to a common rule of measurable comparison.

The purely creative artist is not bound by such rules. Measurement and proportion are unbound myths to be recreated at the artist's whim. But the practical designer, architect, or engineer must be a realist, one who must obey the laws of measurable units, even in the process of visualization.

It is not enough for a building designer to be able to read conversion tables and translate measurements to metric equivalents. Such a person must also equate visually and mentally with metric measure as well. It is a different sort of creativity.

When someone describes a building component in terms of millimeters, a true professional would not ask, "Is it bigger than a bread box?

8
Man Is the Measure

"Man is the measure of all things," observed Pythagoras around 500 B.C. Perhaps he was speaking metaphorically and meant that man's measure governs the design of his environment. If early kings and rulers took this axiom literally, it might explain their preoccupation with arms, hands, thumbs, and feet as instruments of measure. The fact that man's proportions are not uniform caused some incongruities in early systems of measure.

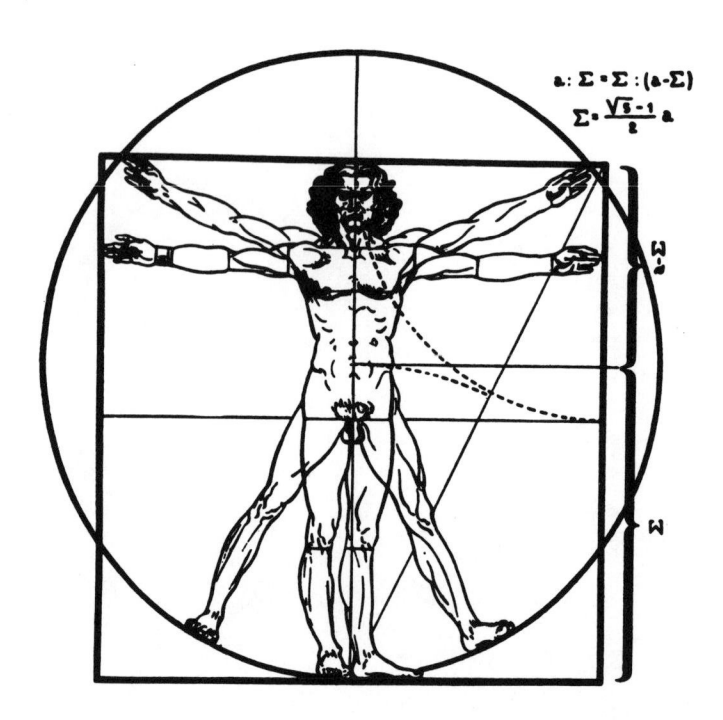

$$\pi\acute{\alpha}\nu\tau\omega\nu \ \chi\rho\eta\mu\acute{\alpha}\tau\omega\nu$$
$$\mu\acute{\epsilon}\tau\rho o\nu \alpha\nu\vartheta\rho\omega\pi o\varsigma$$

$$a : \Sigma = \Sigma : (a-\Sigma)$$
$$\Sigma = \frac{\sqrt{5}-1}{2} a$$

A couple of hundred years after Pythagoras, Euclid defined man's measure in mathematical terms, a formula which became known as the Golden Rule.

In the thirteenth century a young Italian named Fibonacci (who was called "Blockhead" by his neighbors) derived a mathematical sequence of numbers, allegedly by calculating the sequential multiplication of rabbits. In the Fibonacci sequence, each succeeding number equaled the sum of the two preceding numbers, i.e., 1, 2, 3, 5, 8, 13, 21, 34, 55, 89, 144, and so on. It was noted that any number in the sequence divided by the next higher number equaled a constant 0.618034 (accurate from the fourteenth number in the series).

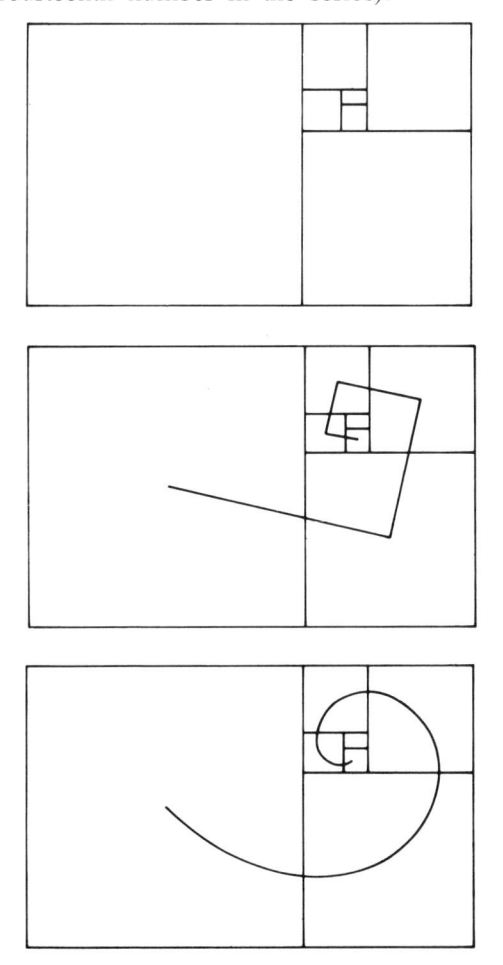

FIBONACCI "THE GOLDEN MEAN"

The magic ratio of 0.618034 to 1 (the Golden Mean) was thereby derived. Graphically the resultant golden rectangle was deemed the perfect proportion. And to most artists and designers it still is. All classic Greek architecture as well as ancient Greek art and sculpture conforms to this ratio. (And there is today an active Fibonacci Association which publishes its own quarterly.) If a square is formed at one end of the rectangle, the division creates another golden rectangle; in turn, if a square is formed within this rectangle another perfect rectangle is created; this process can be repeated ad infinitum, each square and rectangle being smaller than the last. Moreover, if the centers of all squares are connected in a continuous curve, the result is a logarithmic spiral known as the golden spiral, a precise curvature found in many forms of nature such as shellfish and snails. (It is curious that the connotation golden springs up so repeatedly.)

The ancient Greeks adopted the divine proportions as the basis of the measurement of man and proportions of the human figure, and sculptures were created according to these golden ratios. But the secret was lost with the fall of Greece, and not recovered until Leonardo da Vinci recreated the golden proportions in the fifteenth century with his classic diagram of man's proportions.

In the 1930s and 1940s, French architect Le Corbusier pursued the philosophy of Fibonacci's golden mean in considerable detail. If the number sequence and the ratio were a gauge of human proportions, he reasoned, and the golden rectangle was admittedly an ideal architectural proportion as the configuration of classic Greek structures had proven, then the Fibonacci hypothesis could serve as a basis for present-day architectural proportionate solutions, which could both resolve the aim of beauty and accommodate man. Surely, Le Corbusier argued, the metric system was more adaptive to accurate interpretation than were ancient systems of measure.

Despite today's mechanistic technology, building design must still respond to the measure of man.

The study of human factors in dimension and proportion is called *anthropometry*, a vital element in the creation of suitable buildings. Beyond question, man's dimensions affect the design of buildings and their related components. Man is still the measure of his surroundings and environment—the spaces and things that he lives with. Such things relate to his personal comfort and convenience. Architecture is a means of uplifting his environment while at the same time serving man on his own terms. There are, of course, measurable quanta beyond man's physical kinship, measures which are virtually infinite in scope. For example, moonshots and microbes are important to our environment. Nonetheless, in the design of buildings and their furnishings, man's measure is still fundamental.

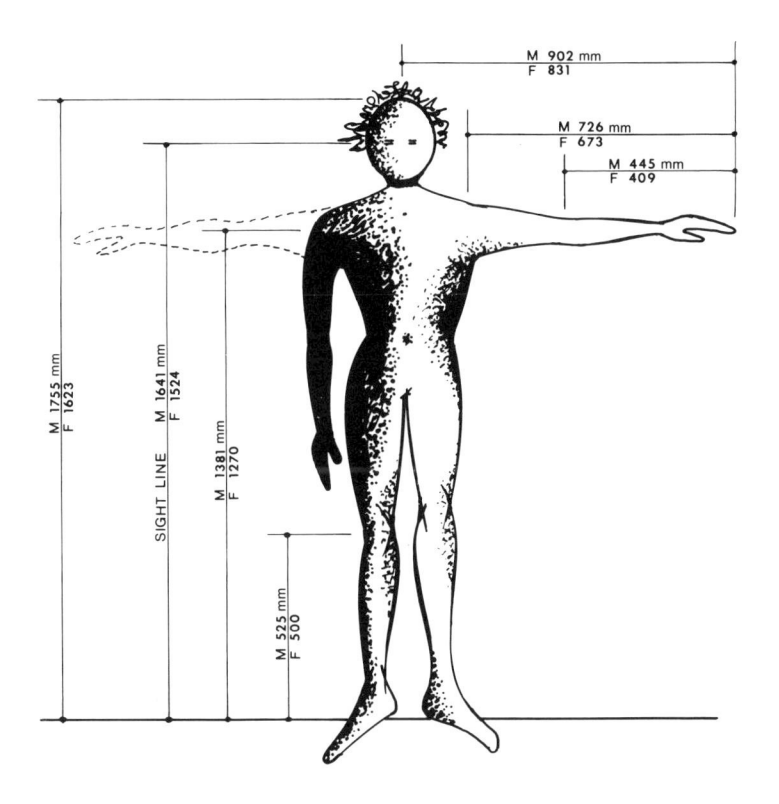

Despite individual differences in human proportions and sizes, the percentiles of both males and females within prescribed age groups remain relatively constant. The chronological growth pattern (height) of normal adults increases approximately 8 mm per decade. Basketball fans may dispute this statistic as being too low.

Because human measurements remain relatively stable, it is doubtful if metric conversions in furniture and human implements will change much in dimension from the English-system equivalents. Seating, for example, may alter dimensions in hard conversion to rounded metric figures, but certainly no more than a few millimeters. Seat height, back height, and width (as well as pitch) are too critical to anthropometric standards and requirements to permit arbitrary change.

THE METRIC PHYSIQUE (MEAN AVERAGES)	in	mm
Standing height		
Male	69.1	1755
Female	63.9	1623
Extended arm length		
Male	28.6	726
Female	26.5	673
Total arm span		
Male	71.0	1803
Female	65.4	1661
Reach (height)		
Male	77.0	1956
Female	71.4	1814
Forward reach (grasp)		
Male	20.2	513
Female	18.7	475
Leg room		
Male	18.0	457
Female	16.3	414
Knee to floor (seated)		
Male	21.7	551
Female	20.0	508
Elbow to fingertips		
Male	17.5	445
Female	16.1	409
Line of sight (standing)		
Male	64.6	1641
Female	60.0	1524
Buttock to knee (seated)		
Male	23.5	597
Female	22.0	559
Shoulder width		
Male	17.7	450
Female	16.0	406
Hip Width (seated)		
Male	13.9	353
Female	14.6	371

Other architectural and furniture products and materials must remain relatively unchanged dimensionally as well, such as desk, work, and counter heights—and to a degree depths. Where accessibility and convenience are the concern of human measure, dimensions must be specified accordingly. Within reason, other prefabricated materials should be in accord with SI preferred planning modules.

Nonmodular dimensions smaller than 100 mm should be in 10 mm increments according to SI recommendations.

The following table gives metric equivalents of typical human dimensions. These are based on mean dimensions, that is, there are an equal number of heights above and below the mean. Thus the measurements given, with their metric equivalents, will serve the greatest number of people A simple average could result in a bias in dimension upward or downward.

In the study of human dimension it has been determined that there are differences in average proportions among different races. Blacks tend to have a longer arm reach and longer legs in proportion to their trunk dimension than do whites; whereas Orientals, in particular the Japanese, have somewhat shorter arm and leg proportions. In certain specialized applications this will affect the metric dimensional choices of furniture and equipment designers. Also it may have a bearing on ergonomic (the relation of people to their environment) considerations, such as working conditions.

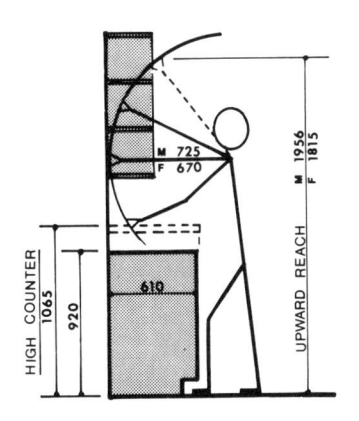

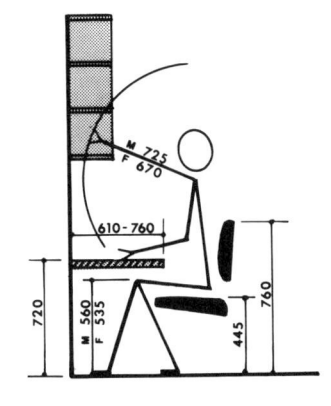

Man's most common positions for living and working are standing, sitting, and lying. Additional postures are bending or leaning, crouching or squatting, kneeling, arching, twisting, and reaching outward and upward. Most tools and implements used by human beings have been engineered and dimensioned to achieve comfort and convenience and can readily be converted to metric measure without significant

56 METRICS FOR ARCHITECTS, DESIGNERS, AND BUILDERS

changes. Today much equipment and furniture—especially in working situations—is adjustable (such as secretarial chairs) in order to provide individual comfort. This was not always true, particularly in period styles of furniture when decorative form and appearance often took precedence over comfort.

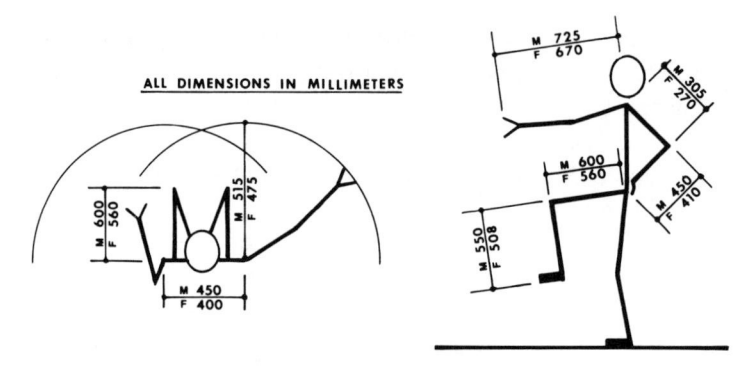

ALL DIMENSIONS IN MILLIMETERS

Over the past several decades there has been considerable debate among designers about the most efficient design of seating. We have become a seated society. We sit to work, we sit to travel, we sit to study and learn, we sit to navigate the golf course or mow the lawn, we even sit to fight our wars. Much of the dimensional engineering in seating has come from European sources and thus is already defined in metric measure.

In the 1940s and 1950s, designers became preoccupied with formed or molded seating shapes in plywood or reinforced fiberglass, reasoning that seat forms that conformed to the contour of the body were more comfortable. The results were successful to a degree, but met with criticism by other designers who insisted that the anthropometrics of the human form could not be so standardized. One critic contemptuously called such molded chair-forms "asstrays." Problems occurred in the manufacture and sale of such contoured seating because many of the original molds were made in Europe, engineered in the metric system, whereas the finished product was to be sold in the United States in English system dimensions and related contours. This is true of other European-made furniture as well. For example, the classic Barcelona chair designed by Mies van der Rohe in Europe measures 750 × 750 × 750 mm, which converts to 29.5 in each dimension. When an American manufacturer began to produce the chair, the dimensions (reverse hard conversion) became 30 in × 30 in × 30 in.

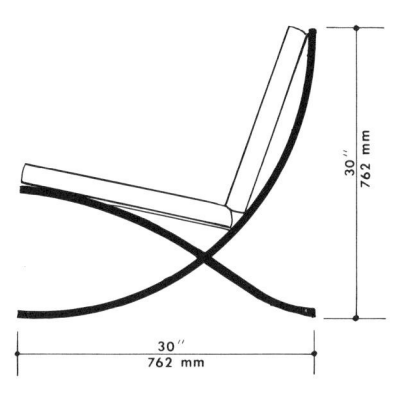

Now much seating—in particular posture-work seating—is designed with adjustable components which are based on metric incremental adjustments.

Other than for seating, there is leeway in adjusting dimensions to comply with human anthropometric requirements, certainly to within 10 mm. Desk heights today vary among manufacturers by as much as an inch, which is a 25.4 mm variable. And workable depths of desks and tables and counters are not critical dimensions.

The important criteria, of course, are the measurements of man.

9
Metrics and the Architectural Environment

It has been said that architecture is to a large extent a matter of selection—not only of products and materials that go into a building design but also of a structural system, a dimensional module, and a design format as well. Standardization of mass-produced and prefabricated materials, including furniture, does simplify, to a degree, dimensional coordination and problems of selection. But in the English system of measure, both dimensional coordination and selection of materials is complicated because of the wide divergence and inconsistency in dimensional standards.

Either much furniture and built-in equipment (such as library book stacks) do not agree with a selected dimensional module or the building module must be adapted to the equipment it must house. Often this is cause for frustration on the part of designer and builder, since mass-produced products cannot easily be altered in size and dimension.

The metric system, with its basic module of 100 mm, will simplify modular dimensional as well as structural coordination. But until total hard conversion is accomplished in all building-related goods and products, it will be necessary to make accurate conversions of all components that go into any metrically dimensioned building design. It can be assumed that building design standards will convert to the metric system before many interrelated mass-produced products, for economic reasons previously mentioned. Such construction materials as masonry, steel, and lumber can be more easily sized and adapted to hard-conversion factors than can mass-produced materials such as paneling, floor tile, or prefabricated doors and windows. For an indeterminate time, there will thus be an overlapping in the conversion process.

Efforts have been made even under the English system to establish standardized building modules. In the 1920s the basic module was ordained as 4 in. In 1934 the National Bureau of Standards attempted to organize dimensional standards. In 1939 the American National Standards Institute (ANSI) initiated further studies in dimensional coordination. By 1957, the Modular Building Standards Association was formed.

Most European nations, already accustomed to the metric system, had adopted the basic 100 mm module by 1961. From this, the International Organization for Standardization (ISO) approved the basic module of 100 mm, with multiples in increments of 25 mm and submodular dimensions of 10, 20, 25, and 50 mm.

Since then lexicographers and other word-wise specialists have struggled over such newly conceived terms such as modularity and metrication.

Any material or article of furniture or equipment that is secured to a building is considered a part of that structure and is thus architectural. This includes not only windows, doors, paneling, floor and ceiling materials, light fixtures, and built-in cabinets, but carpet and wall coverings as well. Other furnishings, accessories, and equipment are called loose furniture.

The conversion to the SI metric system will affect many structural modules and dimensional standards as a result of eventual changes in sizes of building materials and conformity with accepted metric modules. For instance, an 8 ft ceiling height, which equals 2438 mm, will be rounded to 2400 mm (which, conversely, equals 7 ft 10½ in, a decrease of 1½ in).

Metric spacing of studs will be adjusted as follows:

16 in o.c.	which equals 406 mm	will become 400 mm o.c.
24 in o.c.	which equals 610 mm	will become 600 mm o.c.

Such changes will require corresponding changes in the dimensions of applied materials such as paneling and wallboard. A standard drywall or plywood panel which measures 48 in × 96 in (4 × 8) equals 1219 × 2438 mm. This measurement will become, in whole metrics, 1200 × 2400 mm (or 47.2 in × 94.5 in), an adjusted dimension that will conform to the adjusted ceiling height and stud spacing noted above.

Brick sizes will be converted to metric measure: a typical economy brick 4 in × 8 in, which equals 102 × 203 mm, will be rounded to 100 × 200 mm; and a jumbo brick 8 in × 12 in, which converts to 203 × 305 mm, will be adjusted to 200 × 300 mm. Mortar joints will be a standard 10 mm.

In lumber, the dressed or actual dimension of a two-by-four is 1½ in × 3⅝ in, which converts to 38 × 92 mm. This dimension will be rounded to 40 × 90 mm. (Architects and builders will thus no longer speak of two-by-fours but of forty-by-nineties).

It should be pointed out that hard conversion of building material dimensions to metric measure will affect factors of allowable stresses, load factors, elasticity, deflection, and related calculations.

The American National Metric Council has projected 1985 as the date when the construction industry will be totally commited to metric conversion. Since the Metric Conversion Act of 1975 stipulated voluntary, not compulsory, compliance, no deadline is assured. Every fabricator and producer of goods may make his own decision about when he will convert sizes of products to actual whole-metric dimensions. This will occur when the demand justifies the costly change. Hard conversion will require stocking replaceable parts in both present and metric sizes, which will lead to an inflated inventory. Any date that predicts total metric conversion is therefore an assumption.

Building design entails many products and materials, and many different manufacturers. Not all makers of such goods will elect to convert to metric sizes simultaneously; some will employ hard conversion before others. There will be a period when architects must cope with products preconverted to metric sizes as well as products in English system dimensions, which must then be converted to metrics in planning. In effect, there will be a dual system of measure.

For that reason, the data that follow will indicate conversions from the present English system of measure to the metric equivalents.

In the present system of measure, building modules are the decision of the architect, as are building dimensions. There are no prescribed standards. It has been proposed that building dimensions in metrics be broken down as follows:

Up to 3600 mm	300-mm submodules
Up to 9600 mm	600-mm submodules
Over 9600 mm	1200-mm submodules

However, the purpose is to indicate metric conversion equivalents of present or customary dimensions rather than to specify adjusted metric standards as they will exist. This is to clarify more immediate conversion problems to be encountered.

One schedule of building area requirements which has been specified is defined by the Department of Housing and Urban Development (HUD) as minimum residential area standards. These minimum requirements are as shown in the following table, with metric equivalents.

METRICS AND THE ARCHITECTURAL ENVIRONMENT 61

Minimum Living Unit Size

SPACE	1 BR		2 BR		3 BR		4 BR	
	ft²	m²	ft²	m²	ft²	m²	ft²	m²
LR	160	14.864	160	14.864	170	15.794	180	16.723
DR	100	9.290	100	9.290	110	10.219	120	11.148
BR 1	120	11.148	120	11.148	120	11.148	120	11.148
BR (addit)			80	7.432	80	7.432	80	7.432
Total BR's	120	11.148	200	18.581	280	26.013	380	35.303

What follows are common building materials as presently dimensioned with metric equivalents. In some cases millimeter dimensions are carried to one decimal place for accuracy. *This should not be done* in architectural practice. Millimeter notations should be rounded off.

Masonry

Brick sizes have not yet been determined under ISO metric standards. In 1970 the British Brick Development Association (BDA) determined that a single brick size would be standardized at 215 × 102.5 × 65 mm, with a 10-mm mortar joint. In the United States, two brick sizes have been proposed:

Utility brick   100 × 100 × 300 mm (3.9 × 3.9 × 11.8 in)
Standard brick   100 × 67 × 200 mm (3.9 × 2.6 × 7.9 in)

Present brick sizes (all nominal sizes), identified by name, vary. The following table gives standard dimensions with metric equivalents:

	in	mm
Modular	4 × 2-2/3 × 8	102 × 68 × 203
Engineer	4 × 3-1/5 × 8	102 × 81 × 203
Economy	4 × 4 × 8	102 × 102 × 203
Double	4 × 5-1/3 × 8	102 × 135 × 203
Roman	4 × 2 × 12	102 × 51 × 305
Norman	4 × 2-2/3 × 12	102 × 68 × 305
Norman economy	4 × 4 × 12	102 × 102 × 305
Triple	4 × 5-1/3 × 12	102 × 136 × 305
SCR	6 × 2-2/3 × 22	152 × 62 × 305
Norwegian	6 × 3-1/5 × 12	152 × 81 × 305
6 in jumbo	6 × 4 × 12	152 × 102 × 305
8 in jumbo	8 × 4 × 12	203 × 102 × 305

The projected mortar joint thickness will be 10 mm (0.39 in). Present mortar joints in the English system are:

1/4 in	6.4 mm
3/8 in	9.5 mm
1/2 in	12.7 mm

Concrete block size when converted to metrics will be dimensioned 200 × 200 × 400 mm (7.9 in × 7.9 in × 15.7 in). A common standard concrete block is presently 8 in × 8 in × 16 in, although other dimensions are available:

WIDTH		LENGTH		HEIGHT	
in	mm	in	mm	in	mm
2	51	8	203	4	102
4	102	12	305	8	203
6	152	16	406		
8	203				
10	254				
12	305				

Lumber

Wood framing and strcutural members such as joists and rafters can be cut to prescribed length and thus be adapted to a specified metric module or span. Cross-sectional dimensions may remain constant in the conversion to metrics but will eventually be standardized to whole-metric dimensions by the ISO. For the present, studs, joists, and rafters can be used at present cross-sectional dimensions, which are given in the following table (pp. 63 to 65) with metric equivalents.

Flat Panels

For purposes of simplicity in dimensioning, prefabricated panel sizes are listed in two general categories: plywood and drywall, which includes gypsum board, fiberboard, asbestos cement board, particle board, hardboard, etc.

(Continued on page 65)

STRUCTURAL		
NOMINAL, in	ACTUAL, in	METRIC, mm
2 × 4	1-1/2 × 3-5/8	38.1 × 92.1
2 × 6	1-1/2 × 5-1/2	38.1 × 139.7
2 × 8	1-1/2 × 7-1/4	38.1 × 184.2
2 × 10	1-1/2 × 9-1/4	38.1 × 235.0
2 × 12	1-1/2 × 11-1/4	38.1 × 285.8
3 × 6	2-5/8 × 5-1/2	66.7 × 139.7
3 × 8	2-5/8 × 7-1/4	66.7 × 184.2
3 × 10	2-5/8 × 9-1/4	66.7 × 235.0
3 × 12	2-5/8 × 11-1/4	66.7 × 285.8
3 × 14	2-5/8 × 13-1/4	66.7 × 336.6
3 × 16	2-5/8 × 15-1/4	66.7 × 387.4

WOOD COLUMNS		
4 × 4	3-9/16 × 3-9/16	90.5 × 90.5
6 × 6	5-1/2 × 5-1/2	139.7 × 139.7
8 × 8	7-1/2 × 7-1/2	190.5 × 190.5
10 × 10	9-1/2 × 9-1/2	241.3 × 241.3
12 × 12	11-1/2 × 11-1/2	292.1 × 292.1

BOARD THICKNESSES		
1 ×	3/4 ×	19.0 ×
1-1/4 ×	1 ×	25.4 ×
1-1/2 ×	1-1/4 ×	31.8 ×
2 ×	1-1/2 ×	38.1 ×

FLOORING			
	NOMINAL, in	ACTUAL, in	METRIC, mm
Thickness	25/32	25/32	19.8
	1/2	15/32	11.9
	3/8	11/32	8.7
	5/16	10/32	7.9
Face width		1-1/2	38.1
		2	50.8
		2-1/4	57.2

DECKING			
Thickness	2	1-5/8	41.3
	3	2-5/8	66.7
	4	3-1/2	88.9
	5	3-13/16	96.8
Face width	6	5-1/4	133.4
	8	7-1/2	190.5
	10 (laminated)	9-1/2	241.3

	in	mm
PLYWOOD		
Thickness	1/8	3.2
	3/16	4.8
	1/4	6.4
	5/16	7.9
	3/8	9.5
	1/2	12.7
	5/8	15.9
	3/4	19.1
	13/16	20.6
	7/8	22.2
	1	25.4
Width	12	304.8
	16	406.4
	24	609.6
	48	1219.2
Length	48	1219.2
	72	1828.8
	96	2438.4
	144	3657.6
DRYWALL		
Thickness (all types included)	1/12	2.1
	1/10	2.5
	1/8	3.2
	3/16	4.8
	7/32	5.6
	1/4	6.4
	5/16	7.9
	3/8	9.5
	15/32	11.9
	1/2	12.7
	5/8	15.9
	11/16	17.5
	3/4	19.1
Width (all types included)	48	1219.2
	72	1828.8
Length (all types included)	48	1219.2
	60	1524.0
	72	1828.8
	84	2133.6
	96	2438.4
	108	2743.2
	120	3048.0
	144	3657.6

PANEL SIZES

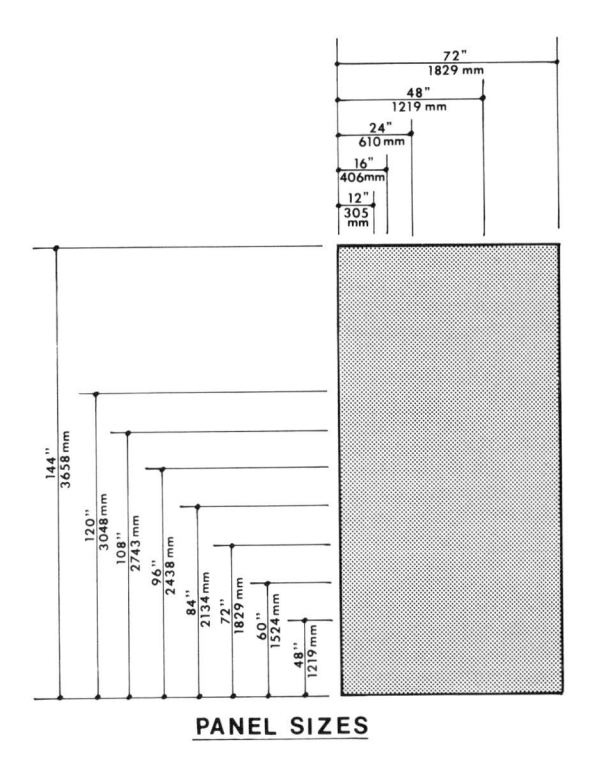

Hard conversion will standardize dimensions to coordinate with accepted metric dimensional modules, i.e., widths: 400, 600, 1200 mm; and lengths: 1800, 2400, 3000, 3600 mm. Present dimensions are as shown in the table on p. 64. (Decimal fractions noted in millimeters are for accuracy, not for use in architectural practice.)

HPL (High-Pressure Laminate)

Known under such trade names as Formica, Wilson Art, Textolite, etc., such plastic sheets are used as surface material for wall paneling and counter and table tops. The material is not rigid and it is bonded to heavier backing materials, normally plywood, chipboard, or hollow metal.

Like most flat panels, the material can be precut or cut on the job, which simplifies dimensional adjustments to metric measure.

Present dimensions are as listed in the table.

	in	mm
Thickness (without backing)		
Heavy duty top-surface wear	1/16	1.6
For postforming (bending)	1/20	1.3
For small radius postforming	1/32	0.8
Width	24	609.6
	30	762.0
	36	914.4
	48	1219.2
	60	1524.0
Length	60	1524.9
	72	1828.8
	84	2133.6
	96	2438.4
	120	3048.0
	144	3657.6

Resilient Flooring

Whether it is in sheet form or in tiles of prescribed dimensions, resilient flooring such as vinyl or vinyl asbestos is trimmed on the job. The size of the area is therefore immaterial. For this reason, the resilient flooring industry may be one of the last to conform to hard conversion of its products. In time, measurements of sheet goods and tiles will be adapted to metric modular standards, which means, however, that in order to calculate area and material requirements in a metrically dimensioned building from flooring catalogued in feet and inches, accurate conversions must be made. Cumulative errors in square meters or millimeters can be costly.

The following are standard dimensions of resilient flooring with metric equivalents.

    Thickness
        1/16 in   1.6 mm
        3/32 in   2.4 mm
        1/8 in   3.2 mm
    Roll width (sheet)
        6 ft   1829 mm
        12 ft   3658 mm
    Tile size
        9 × 9 in   229 × 229 mm
        12 × 12 in   305 × 305 mm

Acoustical Ceiling Panels

Grid framing as well as the dimensions of ceiling panels will be more conveniently installed in metric-dimensioned buildings if they are adapted to metric modules, and will thus provide coordinated modular lighting. This will allow a consistent and articulated grid pattern. Some present common dimensions of ceiling panels are:

Thickness	
1 in	25.4 mm
1-1/2 in	38.1 mm
Panel size	
24 × 24 in	609.6 × 609.6 mm
24 × 48 in	609.6 × 1219.2 mm
30 × 60 in	762.0 × 1524.0 mm
48 × 48 in	1219.2 × 1219.2 mm

Metal

In the category of structural steel shapes such as beams, joists, columns, channels, and angles, there are innumerable standard dimensions. Lengths are as specified and are thus adaptive to any given building module. Typical thickness of cross section dimensions of heavy structural steel shapes are as given in the following table.

in	mm
3/32	2.4
1/8	3.2
3/16	4.8
1.4	6.4
5/16	7.9
3/8	9.5
1/2	12.7
9/16	14.3
LIGHTWEIGHT STEEL BEAMS AND JOISTS (cross-section dimensions)	
3.06 × 12	77.7 × 304.8
2.69 × 10	68.3 × 254.0
2.28 × 8	57.9 × 203.2
2.08 × 7	52.8 × 177.8
1.84 × 6	46.7 × 152.4

68 METRICS FOR ARCHITECTS, DESIGNERS, AND BUILDERS

STEEL ROUND PIPE [inside diameter (nominal)]

1/8	3.2
1/4	6.4
3/8	9.5
1/2	12.7
3/4	19.1
1	25.4
1-1/4	31.8
1-1/2	38.1
2	50.8
2-1/2	63.5
3	76.2
3-1/2	88.9
4	101.6
5	127.0
6	152.4
8	203.2
10	254.0
12	304.8

STEEL SQUARE TUBING (outside dimensions)

1/2 × 1/2	12.7 × 12.7
5/8 × 5/8	15.9 × 15.9
3/4 × 3/4	19.1 × 19.1
7/8 × 7/8	22.2 × 22.2
1 × 1	25.4 × 25.4
1-1/4 × 1-1/4	31.8 × 31.8
1-1/2 × 1-1/2	38.1 × 38.1
1-3/4 × 1-3/4	44.5 × 44.5
2 × 2	50.8 × 50.8
2-1/2 × 2-1/2	63.5 × 63.5
3 × 3	76.2 × 76.2
3-1/2 × 3-1/2	88.9 × 88.9
4 × 4	101.6 × 101.6

STEEL RECTANGULAR TUBING (outside dimensions)

1-1/2 × 1	38.1 × 25.4
2 × 1	50.8 × 25.4
2 × 1-1/4	50.8 × 31.8
2 × 1-1/2	50.8 × 38.1
2-1/2 × 1	63.5 × 25.4
2-1/4 × 1-1/4	57.2 × 31.8
2-1/2 × 1-1/2	63.5 × 38.1
3 × 1	76.2 × 25.4
3 × 1-1/2	76.2 × 38.1

METRICS AND THE ARCHITECTURAL ENVIRONMENT 69

STEEL RECTANGULAR TUBING—*Continued*

in	mm
3 × 2	76.2 × 50.8
4 × 2	101.6 × 50.8
4 × 2-1/2	101.6 × 63.5
4 × 3	101.6 × 76.2
5 × 2	127.0 × 50.8
5 × 2-1/2	127.0 × 63.5
5 × 3	127.0 × 76.2
6 × 2	152.4 × 50.8
6 × 3	152.4 × 76.2
6 × 4	152.4 × 101.6
7 × 5	177.8 × 127.0
8 × 2	203.2 × 50.8
8 × 3	203.2 × 76.2
8 × 4	203.2 × 101.6
8 × 6	203.2 × 152.4
10 × 2	254.0 × 50.8
10 × 4	254.0 × 101.6
10 × 5	254.0 × 127.0
10 × 6	254.0 × 152.4
10 × 8	254.0 × 203.2
12 × 2	304.8 × 50.8
12 × 4	304.8 × 101.6
12 × 6	304.8 × 152.4

FLAT STEEL DIMENSIONS (not all combinations of thickness and width are available)

Thickness		
	1/8	3.2
	3/16	4.8
	1/4	6.4
	5/16	7.9
	3/8	9.5
	7/16	11.1
	1/2	12.7
	9/16	14.3
	5/8	15.9
	11/16	17.5
	3/4	19.1
	7/8	22.2
	15/16	23.8
	1	25.4
	1-1/8	28.6
	1-1/4	31.8
	1-3/8	34.9
	1-1/2	38.1
	1-5/8	41.3

	FLAT STEEL DIMENSIONS—*Continued*	
	in	mm
	1-3/4	44.5
	1-7/8	47.6
	2	50.8
Width	3/8	9.5
	1/2	12.7
	5/8	15.9
	3/4	19.1
	7/8	22.2
	1	25.4
	1-1/8	28.6
	1-1/4	31.8
	1-3/8	34.9
	1-1/2	38.1
	1-5/8	41.3
	1-3/4	44.5
	2	50.8
	2-1/4	57.2
	2-1/2	63.5
	2-3/4	69.9
	3	76.2
	3-1/4	82.6
	3-1/2	88.9
	3-3/4	95.3
	4	101.6
	4-1/4	108.0
	4-1/2	114.3
	5	127.0
	5-1/2	139.7
	6	152.4

Metal Gauges (U.S. Standard Revised)

	in	mm
000	.3750	9.53
00	.3437	8.73
0	.3125	7.94
1	.2812	7.14
2	.2656	6.75
3	.2391	6.07
4	.2242	5.70
5	.2092	5.31
6	.1943	4.94
7	.1793	4.55

METRICS AND THE ARCHITECTURAL ENVIRONMENT 71

METAL GAUGES—*Continued*

	in	mm
8	.1644	4.18
9	.1495	3.80
10	.1345	3.42
11	.1196	3.04
12	.1046	2.66
13	.0897	2.28
14	.0747	1.90
15	.0673	1.71
16	.0598	1.52
17	.0538	1.37
18	.0478	1.21
19	.0418	1.06
20	.0359	0.91
21	.0329	0.84
22	.0299	0.76
23	.0269	0.68
24	.0239	0.61
25	.0209	0.53
26	.0179	0.45
27	.0164	0.42
28	.0149	0.38
29	.0135	0.34
30	.0120	0.30

Flexible Wall Coverings

Wallpaper is sold in single, double, and triple rolls. Widths and roll lengths vary, but regardless of measurements, the combination of width times roll length (trimmed) will equal approximately 36 ft^2 per single roll, which in metric measure is 3.34 m^2. The most common wallpaper measurements are as given in the table.

WIDTH		TRIMMED WIDTH	
in	mm	in	mm
20	508	18	457
27	686	25	635
30	762	28	711
36	914	34	864
41	1041	39	991
48	1219	46	1168

72 METRICS FOR ARCHITECTS, DESIGNERS, AND BUILDERS

SINGLE ROLL LENGTHS		
yd	ft	m
3	9	2.743
5	15	4.572
7	21	6.401

Vinyl wall covering is sold by the linear yard, which equals 914 mm, or 0.914 m. Since the material is produced in rolls, it may be sold by the linear meter, which is 1.094 yd. Standard roll widths are 48 in, or 1219 mm, and 54 in, or 1372 mm.

Some fabric wall covering is sold by the square yard, which equals 0.84 m².

Doors

When hard conversion to metrics is completed, door sizes will be dimensioned in multiples of the basic module of 100 mm. One objective will be to reduce the many sizes now marketed to a minimum number of standard dimensions. It has been suggested, for example, that door heights be standardized at 2100, 2200, and 2400 mm, and widths at 800, 900, 1000, 1200, 1500, and 1800 mm.

Present customary dimensions are as given in the table.

	ft-in	mm
Thickness	1-3/8	34.9
	1-3/4	44.5
Width	1'-6"	457
	1'-8"	508
	1'-10"	559
	2'-0"	810
	2'-2"	660
	2'-4"	711
	2'-6"	762
	2'-8"	813
	2'-10"	864
	3'-0"	914
	3'-4"	1016
	3'-6"	1067

	ft-in	mm
Height	6'–0"	1829
	6'–6"	1981
	6'–8"	2032
	6'–10"	2083
	7'–0"	2134
	7'–6"	2286
	8'–0"	2438

Windows

Differences in design, material, and construction have caused variations in dimensional standards among window types such as casement, single- and double-hung, projected, awning, jalousie, sliding, etc. A distinction is also made between industrial (architectural) and residential window construction and size. Heavy-duty commercial windows demand a different set of structural standards (and thus dimensions) than do lighter-weight residential windows.

Proposals have been made when hard conversion to metric measure occurs, to simplify window sizes to conform to multiples by assembling and joining straight members—extrusions or millwork—hard conversion to metric measure will involve changes in cut lengths of component framing, mullions, etc., and an adjustment of assembly jigs.

The following tables give customary standard dimensions of metal windows, with metric conversions.

CASEMENT AND COMBINATIONS (Fixed and Projected)		
	ft-in	mm
Width	1'–7-1/8"	485.8
	1'–8-1/2"	520.7
	1'–8-7/8"	530.2
	2'–0-1/2"	622.3
	3'–1"	939.8
	3'–4"	1016.0
	3'–4-7/8"	1038.2
	3'–8-1/2"	1130.3
	4'–0"	1219.2
	4'–0-7/8"	1241.4

	CASEMENT AND COMBINATIONS—*Continued*	
	ft-in	mm
	4′–5-1/8″	1349.3
	4′–8″	1422.4
	5′–0-7/8″	1546.2
	5′–9-3/8″	1762.1
	6′–0″	1828.8
	6′–3″	1905.0
	6′–8-7/8″	2054.2
	7′–2-1/4″	2190.8
	7′–7-1/2″	2324.1
	8′–0″	2438.4
Height	2′–1″	636.0
	2′–2″	660.4
	2′–9″	838.2
	3′–1″	939.8
	3′–2-3/8″	974.7
	3′–2-1/2″	977.9
	4′–1″	1244.6
	4′–2-3/8″	1279.5
	4′–2-1/2″	1282.7
	4′–2-5/8″	1285.9
	5′–1″	1549.4
	5′–3″	1600.2
	5′–5″	1651.0
	5′–8″	1727.2
	6′–9″	2057.4
	7′–9″	2362.2
	8′–1″	2463.8
	PROJECTED, SLIDING, AWNING	
Width	1′–7-1/8″	485.8
	1′–8-7/8″	530.2
	2′–0″	609.6
	2′–0-7/8″	631.8
	2′–2″	660.4
	2′–8″	812.8
	2′–8-7/8″	835.0
	3′–0″	914.4
	3′–1″	939.8
	3′–2-3/8″	974.7
	3′–4″	1016.0
	3′–4-7/8″	1038.2
	3′–8″	1117.6
	3′–8-7/8″	1139.8
	4′–0	1219.2

METRICS AND THE ARCHITECTURAL ENVIRONMENT

	PROJECTED, SLIDING, AWNING—*Continued*	
	ft-in	mm
	4′–0-7/8″	1241.4
	4′–2-5/8″	1285.9
	4′–5-1/8″	1349.4
	4′–8-7/8″	1444.6
	5′–0″	1524.0
	5′–0-7/8″	1546.2
	6′–0″	1828.8
	6′–8-7/8″	2054.2
	7′–0″	2133.6
	7′–4-7/8″	2257.4
	8′–0″	2438.4
	8′–0-7/8″	2460.6
	0′–0″	2743.2
	10′–0″	3048.0
Height	1′–2-3/4″	374.7
	1′–5″	431.8
	1′–7-1/8″	485.8
	1′–8-3/4″	527.1
	1′–10-3/4″	577.9
	2′–0″	609.6
	2′–2″	660.4
	2′–8″	812.8
	2′–9″	838.2
	3′–0″	914.4
	3′–1″	939.8
	3′–2-5/8″	981.1
	4′–0″	1219.2
	4′–1″	1244.6
	4′–2-5/8″	1285.8
	4′–5-1/8″	1349.4
	5′–0″	1524.0
	5′–3″	1600.2
	5′–3-3/8″	1609.8
	5′–5″	1651.0
	6′–8″	2032.0
	6′–9″	2057.4
	6′–10-1/2″	2095.5
	7′–0″	2133.6
	7′–5″	2260.6
	8′–0″	2438.4
	8′–1″	2463.8
	8′–2-7/16″	2505.1
	8′–9″	2667.0
	9′–5″	2743.2

	SINGLE AND DOUBLE HUNG	
Width	2′–0″	609.6
	2′–8″	812.8
	3′–0″	914.4
	3′–4″	1016.0
	3′–8″	1117.6
Height	3′–0″	914.4
	3′–8″	1117.6
	4′–4″	1320.8
	5′–0″	1524.0
	6′–0″	1828.8

Note: Millimeter dimensions shown in decimal fractions are for accuracy only. In building-design practice, decimal fractions of millimeters should be omitted and rounded off to whole numbers.

Carpet

Carpet construction is determined by three basic qualities: pile density, pile height, and weight of yarn. Pile density is measured by the following calculations, depending on whether the carpet is woven or tufted.

Pitch, which is the number of yarn loops across the width of a traditionally standard loom of 27 in (686 mm). (This was before broader looms were developed, fostering the term "broadloom." Today the 27-in loom is still used as standard.)

Wires or needles, which represent the number of stitches per inch (25.4 mm) across the loom.

Gauge, or the spacing of needles across the width of the loom.

A typical example:

108 pitch means that there are 108 yarn loops per 27 in (686 mm) loom width.
4 wire means that there are 4 stitches per inch (25.4 mm) laterally across the loom.
1/4 gauge means that there are ¼ in (6.4 mm) between needles.

The question is whether the traditional 27 in (686 mm) will be rounded to 680 mm. If so, 108 pitch will become 107.

Pile height is the height of yarn strands or loops from the backing of the carpet, measured in fractions of an inch. An example might be a 0.218-in pile height, which is 5.5 mm.

Weight of yarn (face weight) means the total weight of face yarn in ounces per square yard. This will undoubtedly become grams per square meter. Yarn count is also a factor. This is the weight of a single strand of yarn measured in linear yards per ounce, or in denier. Oddly enough, denier is a metric measure: the weight of yarn in grams per 9000 m.

These specifications as well as carpet widths will undoubtedly be converted to metric standards, including pricing of carpet by the square yard, or square meter.

Customary carpet widths are as follows:

4'–6 in	1372 mm
6'–0 in	1829 mm
7'–6 in	2286 mm
9'–0 in	2743 mm
12'–0 in	3658 mm
15'–0 in	4572 mm

Total conversion to the metric system of measure will pose both a challenge and a threat to the creativeness of architects. Some will complain of restrictive modular coordination and standardization, which may inhibit freedom of expression. Others will resign themselves to erecting building-block assemblages (which, it must be noted, is often being done today under the English system of measure).

The ancient architect Vitruvius proclaimed the qualities of architecture as strength, utility, and grace. It is the term "grace" that fortifies the antimetric forces. Among foreign architects accustomed to the metric system, however, there is no restraint of imaginative design. Italian architect Nervi, with his meticulously engineered and graceful forms, calculated in metrics.

The advantages of metric modular coordination are obvious. Cost estimating and material inventories will be simplified. Construction schedules will be more precisely controlled. Workers will be more adaptive to simplified modular coordination. Substitution and future replacement of materials and "parts" will be simplified. Use of standard details will simplify drawings, but need not stifle creativity of design. Erection methods and system can be more efficiently organized by utilization of standard components.

By simplified ordering and delivering of materials, building construction will move faster.

However, all this means total conversion to the metric system, of all products and materials relating to the building trade. During the interim period of change—which may take many years—architects and designers must deal with metric conversions of products and materials as they are presently dimensioned.

10
Metrics and the Business Office

Most office furniture manufactured by reputable firms has been carefully engineered and thoroughly tested for functional convenience as well as structural soundness. That is not to say that a nirvana has been reached in the office environment; new improvements and innovations are being introduced constantly by alert firms.

It does mean, however, that hard conversion of business furniture, although costly and time-consuming on the part of manufacturers, will involve relatively little change in actual dimensions. Only a few millimeters added or subtracted will be needed in order to achieve simplified whole-metric measurements. A 30 in × 60 in desk, which equals 762 × 1524 mm, will become 760 × 1520 mm (which conversely equals 29.9 in × 59.8 in).

Because furniture design is based upon anthropometric (human-measurement) standards, its dimensions do not necessarily agree with the 100-mm basic building module, although furniture that is fitted to architecturally dimensioned modules, walls, or partitions—such as storage units—should be adapted to this basic module. For example, the customary widths of lateral files are 30 in, 36 in, and 42 in, which equal 762, 914, and 1067 mm, respectively. With hard conversion these measurements may become 800, 900, and 1000 mm (31.5 in, 35.4 in, and 39.4 in).

Wherever possible, measurements should be converted and correlated with architectural building standards. Interior design drawings and specifications should conform to SI-recommended guidelines of practice for the use of metric units.

Furniture layout will be simplified if spacing follows a metric module pattern, i.e., 36 in aisles between desks or reading tables may be shown as 1000 mm (1 m, or 39.4 in). This adjustment in spacing will change some of the standard formulas used in space planning, such as the customary 125 square feet per work station.

Standard components of business offices, with customary furniture dimensions and metric equivalents, are given in the remainder of this chapter.

Paper

Despite the increased use of computers, data processing, word processing, and information retrieval systems, paper remains a fundamental tool of communication within and without the business office. The federal government still uses tons of paper a day. In the design profession, paper is still the primary means of conveying ideas. Books and periodicals will always be the libraries' stock-in-trade. The mail will continue to deliver paper goods. And typewriters are still valid and vital instruments in office performance.

In the metric system, all paper sizes will be standardized with respect to sheet sizes and proportions. The International Standards Organization (ISO) has recommended what is called its A series, which is basically derived from a rectangle with a total area of 1 m². The ratio of its sides is $1:\sqrt{2}$. The A series begins with a primary dimensional size, identified as A0, which measures 841 × 1189 mm, or 1 m².

The following table shows the recommended (preferred) metric sizes and their equivalents of paper.

SERIES	mm	in
A0	841 × 1189	33.1 × 46.8
A1	594 × 841	23.4 × 33.1
A2	420 × 594	16.5 × 23.4
A3	297 × 420	11.7 × 16.5
A4	210 × 297	8.3 × 11.7
A5	148 × 210	5.8 × 8.3
A6	105 × 148	4.1 × 5.8
A7	74 × 105	2.9 × 4.1

Larger sizes are determined by doubling the shorter dimension and smaller sizes, by halving the larger dimension.

The most common business stationery size will be A4, which approximates the customary 8-1/2 × 11 in. Objections will undoubtedly be heard from the legal profession, which is accustomed to the legal-size 8-1/2 × 14 in. format.

Present customary paper sizes are as follows.

in	mm
3 × 5	76.2 × 127.0
4-1/2 × 7	114.3 × 177.8
5-1/2 × 8-1/2	139.7 × 215.9
8-1/2 × 11	215.9 × 279.4
8-1/2 × 14	215.9 × 355.6
9 × 12	228.6 × 304.8
11 × 17	279.4 × 431.8
12 × 18	304.8 × 457.2
17 × 22	431.8 × 558.8
18 × 24	457.2 × 609.6
22 × 34	558.8 × 863.6
22 × 36	558.8 × 914.4
23 × 36	584.2 × 914.4
24 × 36	609.6 × 914.4
30 × 42	762.0 × 1066.8

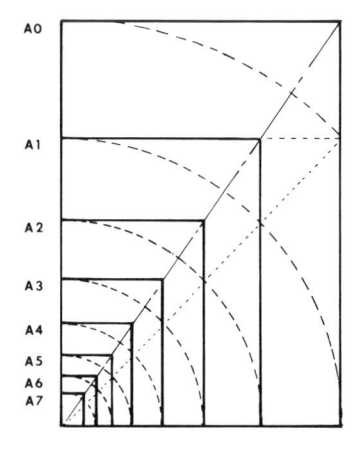

METRIC PAPER SIZES

Office Desks

Often serving as a person's home away from home, the dimensions of a desk are important with respect to function and comfort. Critical

82 METRICS FOR ARCHITECTS, DESIGNERS, AND BUILDERS

dimensions include adequate top size, proper height, comfortable typing height, and sufficient leg and knee space.

Proper desk height has been a subject of controversy in this country. Formerly 30 inches in height, desk heights are now standardized at 29 in, despite the fact that every anthropometric study including the 1944 Hooton studies, the 1959 University of Arkansas studies, and the 1974 MIT *HUMANSCALE 1/2/3* concluded that the most convenient seated working height for the average person should be no more than 28 in. European manufacturers produce desks and worktables at 710 and 720 mm (27.6 and 28.4 in, respectively).

With hard conversion of paper sizes, the dimensions of desk pedestals containing file drawers will undoubtedly be converted accordingly.

The following table lists customary desks as presently dimensioned.

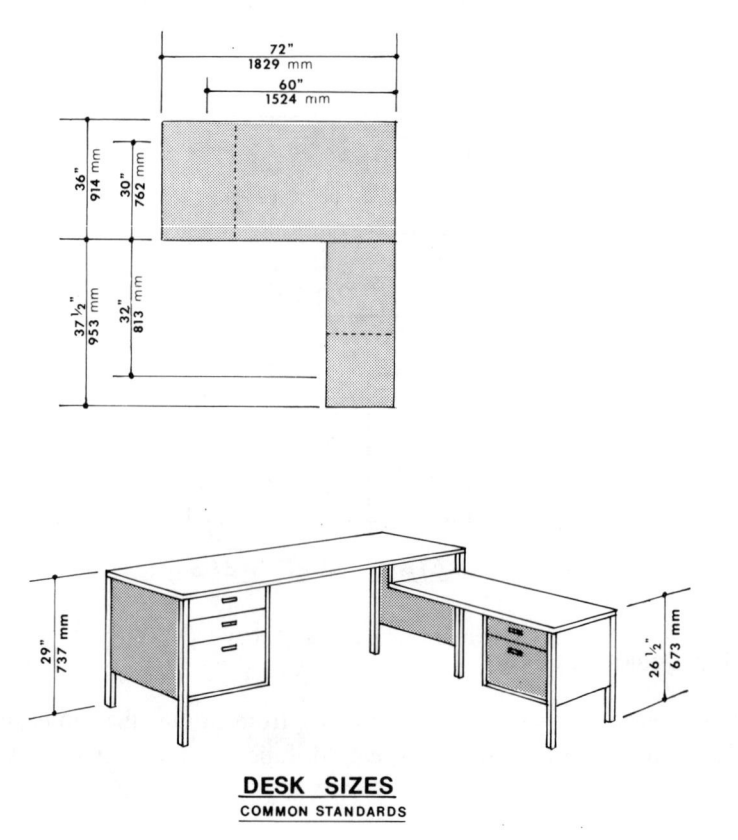

DESK SIZES
COMMON STANDARDS

	in	mm
	TOP SIZES	
Width	24	610
	30	762
	32	813
	36	914
	38	965
	42	1067
Length	48	1219
	60	1524
	66	1676
	72	1829
	78	1981
	84	2134
Height	29	737
	42 (standing)	1067
	RETURNS	
Width	16	406
	18	457
	19-3/4	502
	20	508
Length	23-3/16	600
	30	762
	32	813
	36	914
	37-1/2	953
	42	1067
	45	1143
Height	26	660
	26-7/32	666
	26-1/2	673
	PEDESTALS	
Width	15	381
	15-1/2	394
	16	406
	16-1/2	419
	18	457
Depth	18	457
	19-9/16	497
	20	508
	24	610

Credenzas

Dimensions of credenzas are customarily modular, based on measurements of standard desk pedestals and drawers. Exceptions are credenzas containing doors and shelves. Again, conversion of paper sizes to metric measure will result in dimensional adaptation of file drawers and pedestals and hence changes in the overall dimensions of credenzas.

Present standard dimensions of credenzas are given in the following table.

	in	mm
Width	30	762
	42	1067
	45	1143
	60	1524
	62	1575
	66	1676
	72	1829
	75	1905
Depth	18	457
	19-3/4	502
	20	508
	22	559
	24	610
Height	28-1/2	724
	28-11/16	729
	29	737
	29-3/8	746
	30	762
	31	787
	32	813

Files

Letter- and legal-size files are designed to accommodate paper sizes 8-1/2 × 11 in and 8-1/2 × 14 in. If ISO-recommended paper sizes in metric measure become the preferred standard, the common stationery measurements will be the A4, 210 × 297 mm, which equals 8.3 × 11.7 in. This will require reassessment and redimensioning of filing cases. Adjusted sizes have not yet been determined.

Virtually all business firms and offices have existing files of customary letter- and legal-size documents which cannot be discarded when paper sizes become converted to metric dimensions. It will not be economically feasible to microcopy tons of memos, reports, forms, etc.

Many documents such as original contracts, agreements, or drawings, are rare and irreplaceable in their present form.

For that reason, despite idealistic and hopeful predictions, the hard conversion of standard file dimensions to revised metric sizes will be slow.

FILING CASES

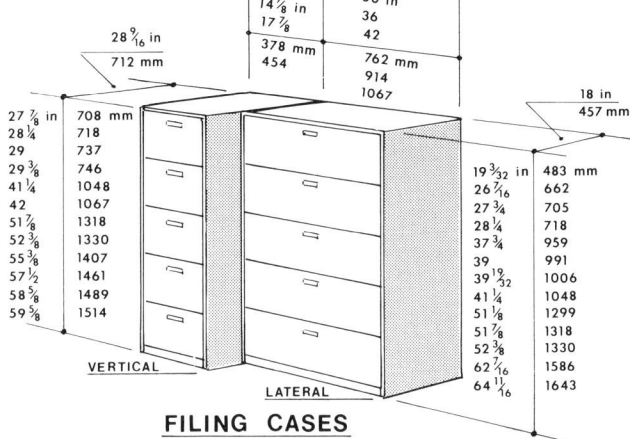

27 7/8 in	708 mm	14 7/8 in	30 in	
28 1/4	718	17 7/8	36	
29	737	28 9/16 in	42	
29 3/8	746	712 mm	378 mm	762 mm
41 1/4	1048		454	914
42	1067			1067
51 7/8	1318	19 3/32 in	483 mm	18 in
52 3/8	1330	26 7/16	662	457 mm
55 3/8	1407	27 3/4	705	
57 1/2	1461	28 1/4	718	
58 5/8	1489	37 3/4	959	
59 5/8	1514	39	991	
VERTICAL		39 19/32	1006	
LATERAL		41 1/4	1048	
		51 1/8	1299	
		51 7/8	1318	
		52 3/8	1330	
		62 7/16	1586	
		64 11/16	1643	

Present legal-size files will accommodate the metric A4 paper; letter-size files will not. Hence firms that manufacture files will undoubtedly market legal-size units in the future. Adaptable inserts will be developed to permit adjustment to the A4 metric dimension of paper. Some files are now marketed which permit adjustment to either customary letter- or legal-size documents.

The dimensions given in the following table indicate present sizes of files, including vertical and lateral files, both letter and legal size, and dimensions of flat files for posters, maps, and drawings.

	in	mm
VERTICAL FILES		
Width	14-7/8	378
	17-7/8	454
	19	483
	20-1/4	514
	21	533
	25-1/8	638
Micro and card	12-7/8	327
Depth	28-9/16	712
Height	27-7/8	708
Two-drawer	28-1/4	718
	29	737
	29-3/8	746

	in	mm
VERTICAL FILES—Continued		
Three-drawer	41-1/4	1048
	42	1067
Four-drawer	50-9/16	1284
	51-7/8	1318
	52-3/8	1330
	55-3/8	1407
Five-drawer	57-1/2	1461
	58-5/8	1489
	59-5/8	1514
LATERAL FILES		
Width	30	762
	36	914
	42	1067
Depth	18	457
Height	26-7/16	662
Two-drawer	27-3/4	705
	28-1/4	718
Three-drawer	37-3/4	959
	38-15/16	989
	39-19/32	1006
	41-1/4	1048
Four-drawer	51-1/8	1299
	51-7/8	1318
	52-3/8	1330
Five-drawer	62-7/16	1586
	63-1/2	1613
	64-11/16	1643
FLAT FILES		
Width	43-5/16	1100
	49-5/16	1253
	55-5/16	1405
	79-5/16	2015
Depth	32-1/2	826
	38-1/2	978
	44-1/2	1130
Height		
Five-drawer	15-3/8	391
Ten-drawer*	15-3/8	391

*Same height as five-drawer file.

Office Seating

All seating dimensions are based on anthropometric proportions aimed at achieving maximum comfort for the purposes for which the seating is intended. Seating measurements, therefore, are derived from human body measurements rather than architectural building modules. Hard conversion to metric measure will be minimal, no more than converting to whole-metric sizes.

Office seating for the performance of office duties or tasks is comprised of the following categories:

Secretarial or clerical
Executive
Conference

Standard seating dimensions among these categories vary because limitation of movement varies. A secretary, clerk, or draftsman is usually limited to certain positions; an executive may alternate from crouching over a desk to assuming a more leisurely posture. The same is true with respect to conference seating.

Most studies of seat dimensions (overall width, depth, height, seat height, and armrest measurements when they exist) give minimum and maximum sizes to accommodate different body measurements and an average preferred mean dimension.

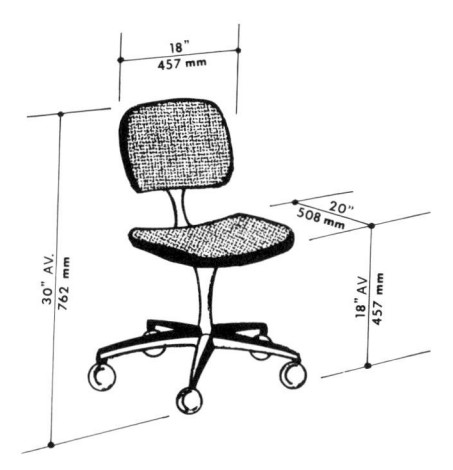

The following tables give office-seating measurements by categories with metric equivalents.

88 METRICS FOR ARCHITECTS, DESIGNERS, AND BUILDERS

	CUSTOMARY, in		METRIC, mm		OPTIMUM, in	METRIC, mm
	MINIMUM	MAXIMUM	MINIMUM	MAXIMUM		
SECRETARIAL						
Width	16	19	406	483	18	457
Depth	18	22	457	559	19	483
Height	27	34	686	864	29.5	749
Seat height*	16	20	381	457	17.5	445
EXECUTIVE						
Width	24	31	610	787	None	None
Depth	24	29.5	610	749	None	None
Height	32	48.5†	813	1232	None	None
Seat height	15	20	381	508	18	457
Seat Depth	17	19.5	432	495	18	457
Armrest height‡ (from Seat)	7	10	178	254	8.5	216
CONFERENCE						
Width	22	32	559	813	None	None
Depth	22	30	559	762	None	None
Height	29	39	737	991	None	None
Seat height	17	20	432	508	17.5	445
Armrest height (from seat)	7	10	178	254	8.5	216
SWIVEL STOOLS						
Width	18	24.5§	457	622	18.75	476
Depth	18.5	26	470	660	20	508
Height	35.5	42	902	1067	None	None
Seat height	25	32	635	813	Adjustable	Adjustable

†Includes headrest.
*Adjustable seat heights from 16 to 19-3/4 in (406 to 502 mm). Center of back lumbar support should be from 9 to 10 in above seat (229 to 254 mm). No restrictive back support should be provided below the sacrum, initial 3 in above seat (76 mm).
‡Armrest clearance beneath work surface top should be a minimum of 1.5 in or 38 mm. Clearance between armrests should be 19 to 22 in (483 to 559 mm).
§With armrests.

Most office seating today provides a number of adjustments, especially in secretarial chairs. This includes adjustable seat and back heights, and in some cases the depth from the front of the seat to the adjusted back, tilt tension control, and swivel return. Adjustments are made to accommodate users of different sizes and shapes. Each adjustment cannot be measured in millimeters: the dimensions in the table are therefore generalizations based on manufacturers' standards.

METRICS AND THE BUSINESS OFFICE 89

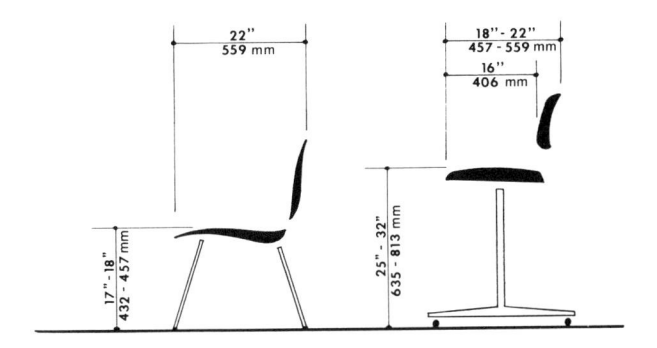

Conference Tables

There are no structural or functional factors in the design of conference tables which will inhibit eventual hard conversion to metric measure. Furthermore, conference tables are free-standing entities and need not adapt to building design modules. Revised sizes will simply be adapted to whole-metric dimensions. The timing of such conversion to metric dimensions will be up to the manufacturers.

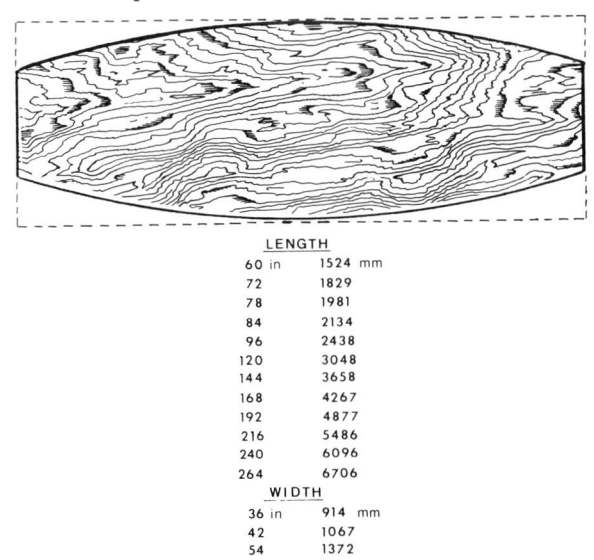

LENGTH	
60 in	1524 mm
72	1829
78	1981
84	2134
96	2438
120	3048
144	3658
168	4267
192	4877
216	5486
240	6096
264	6706

WIDTH	
36 in	914 mm
42	1067
54	1372

CONFERENCE TABLES

Seating-allowance width for conference use should be a minimum of 27 in (686 mm) per seat.

The following table enumerates customary dimensions of both rectangular and boat-shaped conference tables.

	in	mm
RECTANGULAR TABLES		
Length	60	1524
	72	1829
	78	1981
	84	2134
	96	2438
	120	3048
	144	3658
	168	4267
	192	4877
	216	5486
	240	6096
	264	6706
Width	36	914
	42	1067
	48	1219
	54	1372
	60	1524
Height	28-1/2	724
	29	737
BOAT-SHAPED TABLES		
Length	60	1524
	72	1829
	78	1981
	84	2134
	96	2438
	120	3048
	144	3658
	168	4267
	192	4877
	216	5486
	240	6096
	264	6706
Width	42/32	1067/813
	48/36	1219/914
	54/40	1372/1016
	60/42	1524/1067
	66/48	1676/1219
	72/54	1829/1372
Height	28-1/2	724
	29	737

*Width dimensions of boat-shaped tables show both the center and end measurements.

METRICS AND THE BUSINESS OFFICE   91

Movable Wall Systems

The concept of movable partitions which can be assembled and disassembled without attachment to the structure of a building has become a standard tool of interior planning. Many manufacturers offer such systems constructed of a variety of materials, connected or joined by various methods and incorporating a diversity of working components and attachments. The systems are in themselves modular.

Movable wall systems are designed to correlate with building dimensional modules. Although not a part of the structure, modularity of movable wall panels must be compatible with dimensional and structural modules. As building design turns to metric measure, so must the dimensional module of movable wall panels become metric. Hard conversion of movable wall systems will undoubtedly parallel the architectural transition to metric measure. The basic 100-mm building module will prevail.

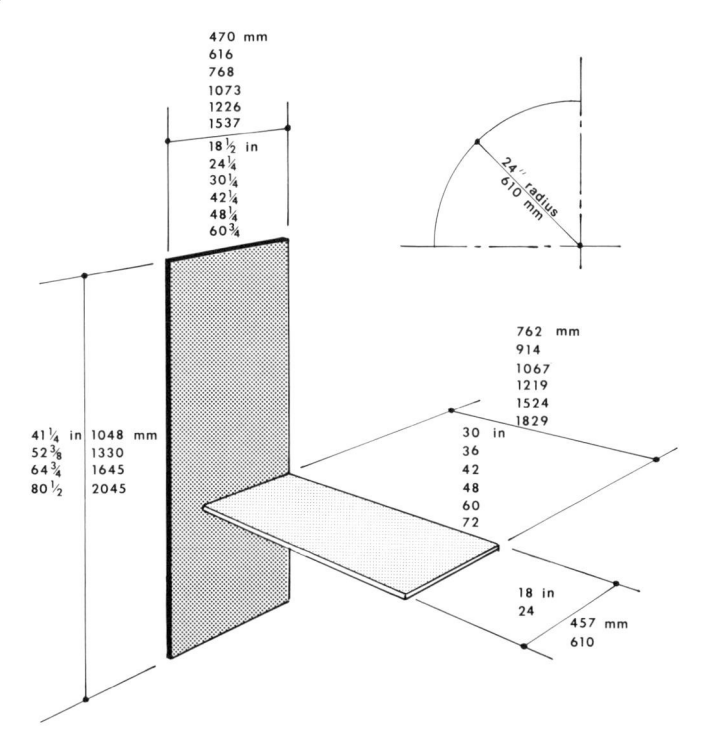

470 mm
616
768
1073
1226
1537

18½ in
24¼
30¼
42¼
48¼
60¾

24 in radius
610 mm

762 mm
914
1067
1219
1524
1829

41¼ in   1048 mm
52⅜   1330
64¾   1645
80½   2045

30 in
36
42
48
60
72

18 in
24

457 mm
610

PANEL SYSTEMS

92 METRICS FOR ARCHITECTS, DESIGNERS, AND BUILDERS

The table that follows gives present customary dimensions of movable wall panels, work-surface attachments, and component units. Note that dimensions of file pedestals and bins will accord with changes in paper sizes to metric measure to accommodate new paper sizes.

Also note that vertical hanging heights of attached components, customarily designated as 1 in o.c., will become a metric integer, perhaps 25 mm o.c. This will simplify the metric-adjusted desk, typing, and working-surface heights.

	in	mm
PANELS		
Width	12	305
	18-1/2	470
	24-1/4	616
	30-1/4	768
	39	991
	42-1/4	1073
	48-1/4	1226
	60-3/4	1537
	62	1575
Height	41-1/4	1048
	42	1069
	48-1/4	1226
	52-3/8	1330
	62-1/4	1581
	64-3/4	1645
	80-1/2	2045
Thickness	1-1/4	32
	2	51
Curved (radius)	24	610
WORK-TOP SURFACES		
Span width	30	762
	36	914
	42	1067
	48	1219
	60	1524
	72	1829
Depth	18	457
	24	610
	30	762

METRICS AND THE BUSINESS OFFICE 93

Work-surface area should be a minimum of 36 in wide × 24 in deep (864 in², which equals 557 418 mm²), or 914 × 610 mm.

Depth of typing stands should be 16 in to 18 in, or 406 to 457 mm. Leg clearance beneath typing stands should be no less than 16 in, or 406 mm, for the average adult.

	in	mm
PEDESTALS AND DRAWER UNITS		
Width	15-1/2	394
	18-1/4	464
Depth	15-3/8	391
	20	508
	22-3/8	568
	24	610
Height	3	76
	11	279
	15-5/8	397
	18-3/8	467
	24-3/8	619
SHELVING AND FILE UNITS		
Width	24	610
	30	762
	36	914
	42	1067
	48	1219
	60	1524
Depth	12	305
	12-1/2	318
	13	330
	14-5/8	372
	15	381
	15-1/2	394
	16	406
	20	508
Height	7-1/2	191
	11-7/8	302
	12	305
	13-1/2	343
	15-1/2	394
	16-7/8	429
	17-1/2	445

Movable wall systems allow adjustable heights of hanging components on 1 in vertical centers. For the average adult, work-surface heights should be from 27 in to 28.5 in, or 686 to 724 mm. Typing-surface heights should be from 25 to 26.5 in, or 635 to 673 mm. The advantage of adjustability of component heights is that each user can determine the most convenient and comfortable working height.

Standing-height work surfaces should be adjusted from 36 to 42 in, or 914 to 1067 mm, depending on the height and arm reach of the user. Depth should be 18 to 24 in (457 to 610 mm).

Hard conversion of movable panels and work components to metrics will undoubtedly be correlated with preferred building modules. Dimensional coordination will make movable wall systems compatible with overall building design and construction and therefore will simplify planning as well as installation.

Recommended architectural modules, which are multiples of the basic 100 mm module, are:

$$
\begin{aligned}
300 \text{ mm} &= 11.8 \text{ in} \\
600 \text{ mm} &= 23.6 \text{ in} \\
900 \text{ mm} &= 35.4 \text{ in} \\
1200 \text{ mm} &= 47.2 \text{ in} \\
1500 \text{ mm} &= 59.0 \text{ in} \\
3000 \text{ mm} &= 118.1 \text{ in} \\
6000 \text{ mm} &= 236.2 \text{ in}
\end{aligned}
$$

The equivalents in inches are given as reference. It is apparent that the multimodules listed above are insufficient to meet the total requirements of a movable wall system. For such a system to be workable, a multimodule of 700 mm (27.6 in) or 800 mm (31.5 in) must be included, as well as 1800 mm (70.9 in). These will satisfy present needs of 30 in for work-surface depths, and 72 in for other dimensional needs. Also a multimodule of 2400 mm (94.5 in) will resolve certain panel height requirements.

The fact that heights of attached component units are adjustable, presently at 1 in intervals, which may be converted to 25 mm (0.98 in), will resolve the anthropometric measurement relationships to seating and working, e.g., panel dimensions will agree with modular building standards but working heights of components will conform to human sizes. Comfortable working conditions are sometimes at odds with agreeable working environments.

11
Metrics In Restaurants and Bars

We have become fast-food addicts. The old-fashioned virtues of family-style restaurant dining with five or six kids around a massive table are virtually gone, being supplanted by take-home service and the TV. Restaurant lunch is for business people and social shoppers ("I'll meet you at Timbuctu's"); dinner is adult-oriented and oftentimes serves as a background for a business deal. This has affected restaurant layout in two ways: tables are uniformly smaller, and seating plans are geared to two- and four-person accommodations.

In drinking establishments the bar is the focal point, easily accessible and often spanning two sides of the barroom. The length of bars is therefore a variable dimension.

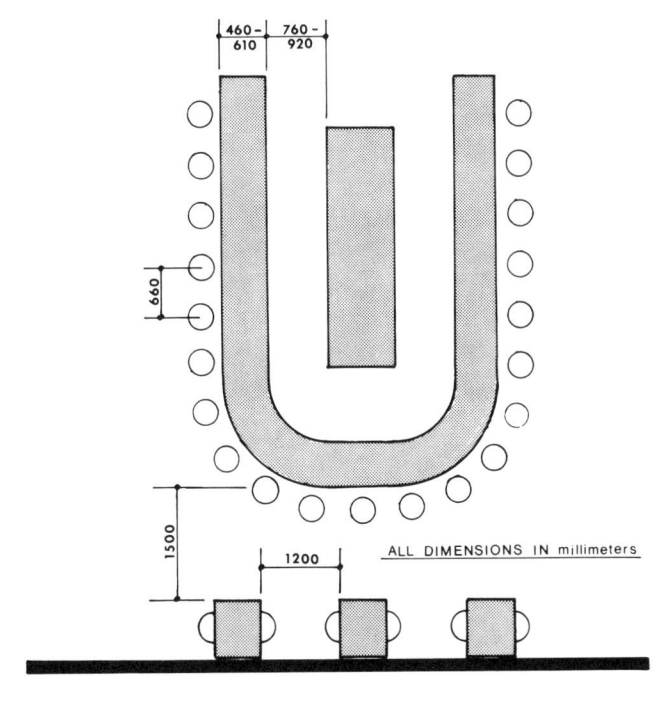

Restaurant and bar tabletop dimensions with metric equivalents and recommended whole-metric conversions; are given in the following table.

	TOP SIZE	
in	mm	RECOMMENDED mm
20 × 20	508 × 508	510 × 510
20 × 24	508 × 610	510 × 610
24 × 24	610 × 610	610 × 610
24 × 30	610 × 762	610 × 760
24 × 42	610 × 1067	610 × 1060
30 × 30	762 × 762	760 × 760
42 × 42	1067 × 1067	1060 × 1060
40 (diameter)	1016 (diameter)	1000 (diameter)
42	1067	1000

Some bars provide small two-drink tables that are customarily 18 × 18 in (457 × 457 mm).

The depth of bar and counter tops should be deep enough for service but not so deep as to hinder service by the bartender. Typical bar-counter depths are:

| Lunch counters | 18 to 30 in | 457 to 762 mm |
| Bar service | 18 to 24 in | 457 to 610 mm |

The height of dining tables is customarily 29 in, or 737 mm, although a lower height of no more than 28.3 in (720 mm) is recommended. Bar-counter heights vary from 36 in to 45 in (914 to 1143 mm); recommended heights are 920, 1020, and 1120 mm, which equals 36.2, 40,2, and 44.1 in, respectively.

Spacing of tables is important. Proprietors of restaurants and bars seek a happy medium between satisfying customers with uncrowded conditions and providing maximum seating for maximum turnover at the cash register.

Clearances between tables should be adequate for easy customer access and seating without interference with other seated customers. Ample intercrossing service aisles are vital to efficient restaurant service. Clearance for such aisles should be no less than 900 mm (35.4 in) and preferably as wide as 1100 mm (43.3).

Table seating will be discussed in detail in Chapter 12, under library seating; but the proportions of dining chairs are not as critical as are library chairs because dining is a single function. A minimum of postural variations are required. Following are accepted customary dimensions of restaurant dining chairs:

Width of seat	16-1/2 in	419 mm
	17-1/2 in	445 mm
Height of seat	17 in	432 mm
	18 in	457 mm
Depth of seat	15-1/2 in	394 mm
	16-1/2 in	419 mm
Width of back	12 in	305 mm
	14 in	356 mm

Some luxury restaurants provide larger fully upholstered seating (with correspondingly greater spacing between tables). However, the practical requirements of restaurant planning call for maximum seating within the allocated space. Again, restaurateurs strive for that apt combination of uncrowded conditions and maximum customer accommodations.

The following charts and diagrams show customary restaurant arrangements and spacing. Where possible, measurements have been rounded to whole-metric dimensions.

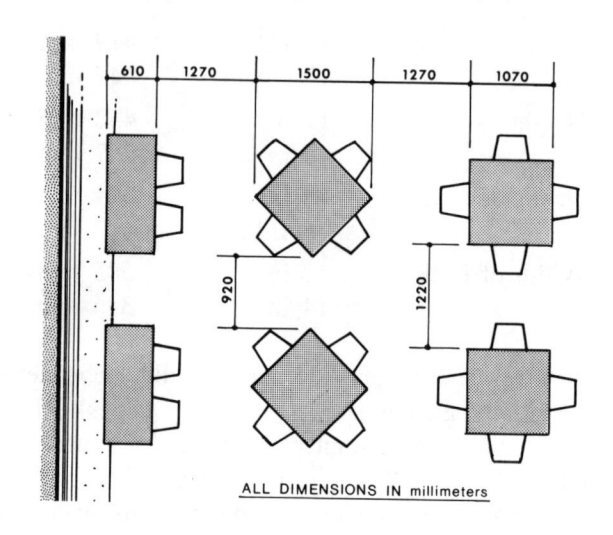

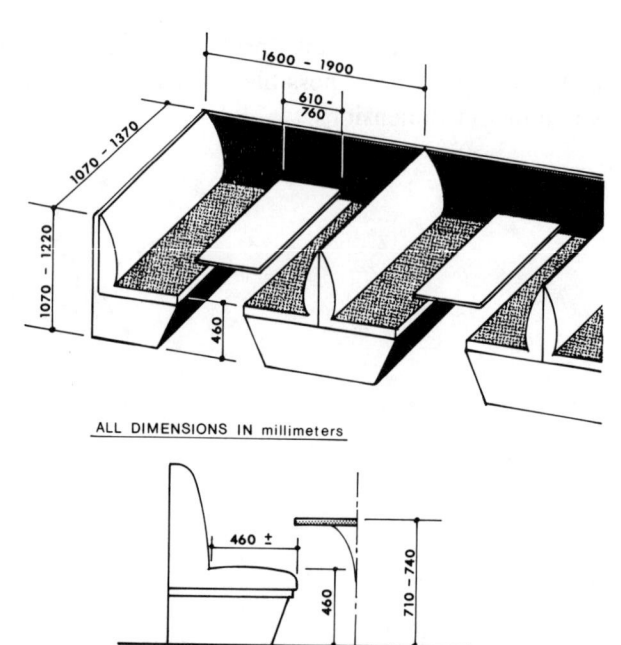

98 METRICS FOR ARCHITECTS, DESIGNERS, AND BUILDERS

ALL DIMENSIONS IN millimeters

ALL DIMENSIONS IN millimeters

12
Metrics and Libraries

Shelving

No building design will be affected more by our conversion to the metric system than will be that of libraries, primarily because the arrangement and spacing of book stacks must be coordinated with building modules. Ranges of shelving must relate to the modular spacing of interior columns.

Architects must often adapt their building modules to the fixed dimensions of shelving sections and the allowable limits of spacing between stack ranges in order to achieve optimum storage capacity for books and related materials. In areas containing multiple stack ranges. spacing is governed by the dimension of the structural bay module. In building design, conversion to metric measure will influence the spacing of stacks and possibly the number of ranges per bay.

The most common type of shelving used in libraries is the bracket type, which consists of vertical uprights which bear the load of cantilevered shelves secured to the uprights by means of lugs. The distance between uprights measured center to center constitutes a section of shelving. This dimension and that of the workable distance between ranges are the factors that determine an efficient bay module.

In his reference volume, *Planning Academic and Research Library Buildings* (McGraw-Hill, 1965), Keyes D. Metcalf examines in detail modular planning relative to book stacks. His calculations are based on current customary measurements. No structural bay module is ideal for planning book stacks, concludes Metcalf, but "the most workable bay modules relating to stack dimensions and acceptable aisle widths are 22 ft.–6 in. × 22 ft.–6 in., and 25 ft.–6 in. × 25 ft.–6 in., with bracket-type shelving." Metric conversions are as follows:

$$22'\text{-}6'' \times 22'\text{-}6'' = 6858 \times 6858 \text{ mm}$$
$$25'\text{-}6'' \times 25'\text{-}6'' = 7772 \times 7772 \text{ mm}$$

The customary length of a section of shelving is 36 in (914 mm). Hard conversion of 914 mm to whole-metric lengths poses no apparent engineering or fabricating problem. The length of a section is determined by the length of the shelves and the length of the top and bottom crossrails or stringers. These members are simply formed and their length is easily adjusted to the manufacturing process.

Undoubtedly the section-length dimension of 914 mm will be rounded to 900 mm, which is common in most foreign nations. This conforms to a multiple of the basic building module of 100 mm. In metric building practice where multiples of the basic 100-mm module are recommended, one suggested multimodule is 6000 mm, because this figure is divisible by 200, 300, 400, 500, 600, 1000, 1200, 1500, 2000, and 3000 mm. Intermediate modules are acceptable in planning, but submultiples of 100 mm are to be avoided unless necessary.

The length of stack ranges should not exceed 10 sections and preferably not more than 7 sections without intermediate cross aisles. Thus the following multiples (section lengths, in millimeters):

$$900$$
$$1800$$
$$2700$$
$$3600$$
$$4500$$
$$5400$$
$$6300$$
$$7200$$
$$8100$$
$$9000$$

Spacing of ranges, i.e., the distance apart measured on centers, is the sum of the aisle width plus the depth of the stack. Three conditions affect this dimension:

1. Closed stack areas, which are not accessible to the public or library users, require no more than minimum access aisles.
2. Open stack areas, which are accessible to patrons as well as library staff, demand wider aisles to handle user traffic.
3. Handicapped code requirements, now in effect in many states, call for aisle clearances ample for passage and turning of wheelchairs. Code requirements specify a minimum aisle width of 42 in (1067 mm).

Metric conversions of customary aisle widths are as follows:

DESCRIPTION	MINIMUM		PREFERRED	
	in	mm	in	mm
Closed stacks	23	584	27	686
Open stacks	30	762	36	914
Handicapped code	42	1067		

There is considerable dispute about the practicality of wider aisles required by handicapped codes as opposed to assistance from library staff members. For instance, the upward reach from a wheelchair averages 59 in (1499 mm) and the downward reach is some 13 in (330 mm) from the floor; many library shelves are therefore inaccessible to the chairbound. Furthermore, the allocation of excessive and valuable floor space is far more costly than is occasional help from a staff member. Buildings are valued by overall usable space.

Shelf depths vary to accommodate different-sized books and material. Thus, 8 in shelves will house more than 90 percent of all books; 10 in and 12 in shelves are specified for large volumes such as art books, children's picture books, and reference volumes; deeper shelves are used for oversized material such as newspapers (folded) or flat-shelved atlases. Nominal dimensions of shelf depths are:

in	mm
8	203
10	254
12	305
16	406
18	457

In planning a series of aligned ranges a vital dimension is the *actual* depth of double-faced shelving, which is a greater dimension than shelf depths because of a deeper base. The customary overall depth is 24 in (610 mm).

- The stack depth of 610 mm plus a workable aisle width equals the spacing of ranges.
- Multiples of range intervals should correlate with multiples of

900-mm section lengths to achieve a uniformly square structural module.

The illustrations indicate some possible solutions.

(A) Column size is an unknown factor, but the illustrations show columns no larger than one section of shelving (900 × 610 mm, or 35.4 in × 24 in). A modular span of 7200 mm provides ranges 9 sections long, with columns absorbed at the ends of two ranges. Spacing of ranges prorates to 6 ranges and 5 aisles, at 1440 mm on centers.

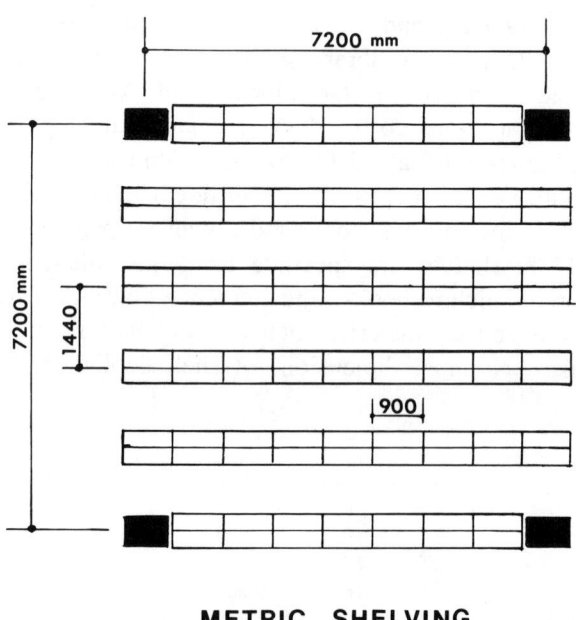

METRIC SHELVING
A

(B) No combination of multiple sections and spacing allowances will satisfy, even closely, the handicapped code requirement of 1067-mm aisle clearance and achieve a square bay module. The illustration shows the closest apparent solution, a range spacing of 1800 mm, which results in a wasteful aisle width of 1190 mm, or 46.9 in, considerably over the minimum required by handicapped codes.

METRIC SHELVING
B

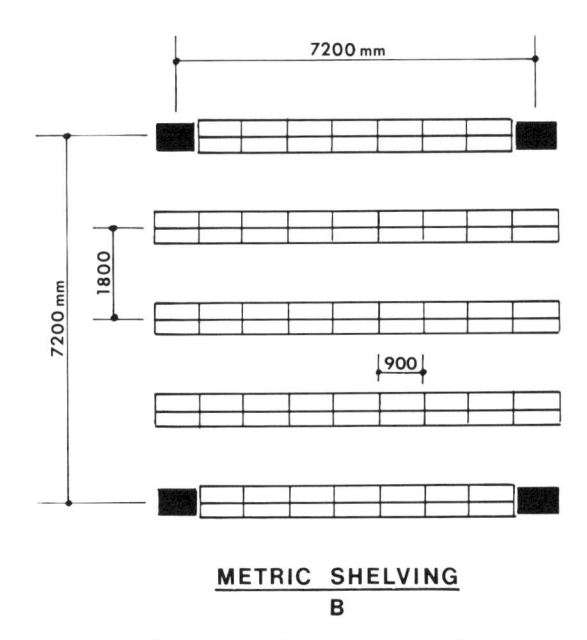

(C) This plan offers a possible solution when bay modules do not conform to multiples of the 900-mm section length. In all cases, the spacing of ranges should create a usable but not unnecessarily wide aisle.

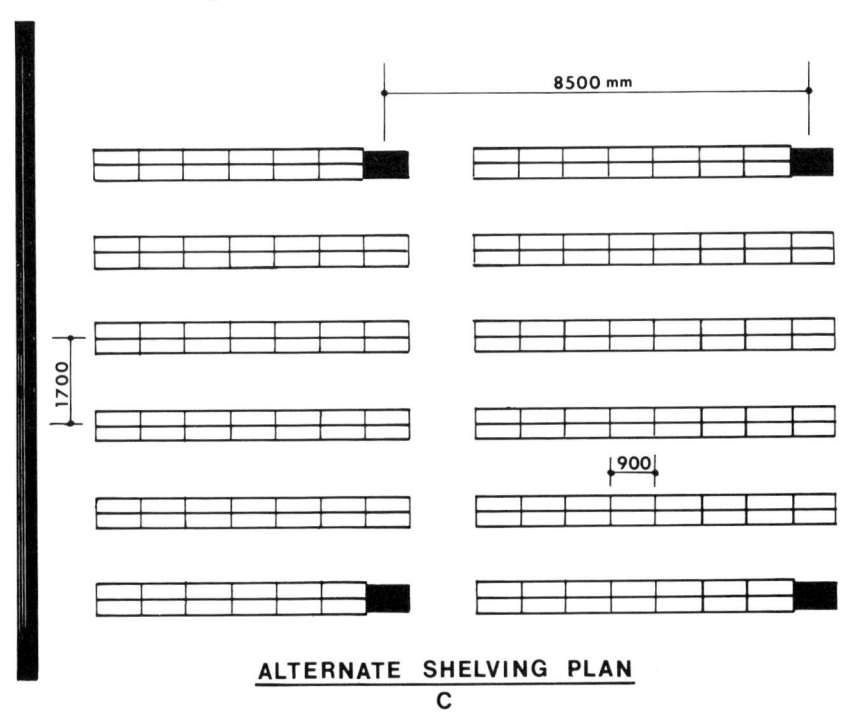

ALTERNATE SHELVING PLAN
C

In summary, a spacing of ranges at 1440 mm on centers is adequate for both closed and open stacks; or 1700 mm where the minimum handicapped code requirement must be complied with.

A new library building must oftentimes incorporate ranges of existing shelving which are still usable. This may involve planning book stacks of 914 mm (36 in) sections in a metrically dimensioned building. Also it may entail combining section dimensions, e.g., existing sections at 914 mm with newly procured shelving at 900 mm per section.

Based on the present customary section length of 914 mm, the following are multiple section lengths in metric dimensions:

NUMBER OF SECTIONS	mm
1	914
2	1828
3	2742
4	3656
5	4570
6	5484
7	6398
8	7312
9	8226
10	9140

The *precise* metric equivalent of 36 in is 914.4 mm. The cumulative error in 10 sections over 900-mm section lengths is therefore 144 mm.

Customary heights of book shelving have been standardized in accordance with the needed clearance between shelves, to accommodate book dimensions. Standard heights of shelving are:

in	mm
42*	1067
48	1219
60	1524
66*	1676
72	1829
78	1981
84	2134
90*	2286

*Most commonly used.

These heights will accommodate anywhere from three to seven shelves vertically. Of the three most commonly used heights noted above, 1067 mm may be rounded to 1100 mm; 1676 mm to 1700 mm; and 2286 mm to 2300 mm.

Card Catalogs

The traditional card catalog is likely on the way out, to be replaced by the more sophisticated Computer Output Microfilm (COM) units; but many libraries will have to continue with the card file system owing to the cost of retrospective conversion of their bibliographic record to machine-readable format. Because of the high cost of changeover the card catalog will likely remain in many libraries.

Standard catalog card size is 3 in × 5 in, or 76 x 127 mm. If the ISO recommended metric standard for paper sizes is adopted, the nearest A size to 3 in × 5 in will be the A7, which is 74 × 105 mm, or 2.9 in × 4.1 in. This transition, at least in catalog cards, is doubtful simply because sturdy and costly card catalog cases are made to last often as long as 20 to 30 years. Catalog drawers are constructed precisely to accommodate 3 in × 5 in cards. Ditching $2000 card catalog cabinets would be prohibitively expensive.

Following are customary dimensions of card catalog cases with metric equivalents:

5 Drawer (case only)
33-1/2" W × 18" D × 5-1/2" H   851 × 457 × 140 mm

10 Drawer (case only)
33-1/2" W × 18" D × 9" H   851 × 457 × 229 mm

15 Drawer (case only)
33-1/2" W × 18" D × 13" H   851 × 457 × 330 mm

Leg Base Height
16" H   406 mm
26" H   660

60 Drawer (overall dimension)
40-1/2" W × 18" D × 60" H   1029 × 457 × 1524 mm

72 Drawer (overall dimension)
40-1/2" W × 18" D × 64-1/4" H   1029 × 457 × 1632 mm

Service Desks

Library service desks provide a key function of library operation. Their purpose is to circulate materials, furnish reference and information, and maintain control of material withdrawal and return. Following are customary dimensions of prefabricated units:

	in	mm
Width (component)	30	762
	36	914
Depth	24	610
	27	686
	30	762
Height (standing)	39	991
	42	1067

There is, however, a preference among many architects and interior designers to design and detail service desks according to the particular needs of the library. This permits the desk to be adapted precisely and individually to the job.

Carrels

The top surface dimensions of individual study carrels should be no less than 36 in w × 24 in d (914 × 610 mm). In college or university libraries and for audiovisual units, carrels are sometimes 48 in (1219-mm) wide.

Carrels are often assembled in a variety of configurations using the above basic dimensions as the unit size.

Single carrels which are aligned, such as those abutting a wall or partition, should allow minimum spacing of 1400 mm (55 in) on centers which, with a carrel depth of 610 mm, permits clearance of 790 mm (31 in). Aisle clearance at ends of carrels should be 1000 to 1100 mm (39.4 to 43.3 in) in order to accommodate wheelchair passage for the handicapped. With seating on either side, clearance between carrels should be no less than 1676 mm (66 in) to allow for the double flanks of seating.

The width of a wheelchair averages 686 mm (27 in). Therefore, for straight passage (without turning allowance), a minimum clearance between any obstacles should be 770 mm (30.3 in).

CARRELS

CARREL CONFIGURATIONS

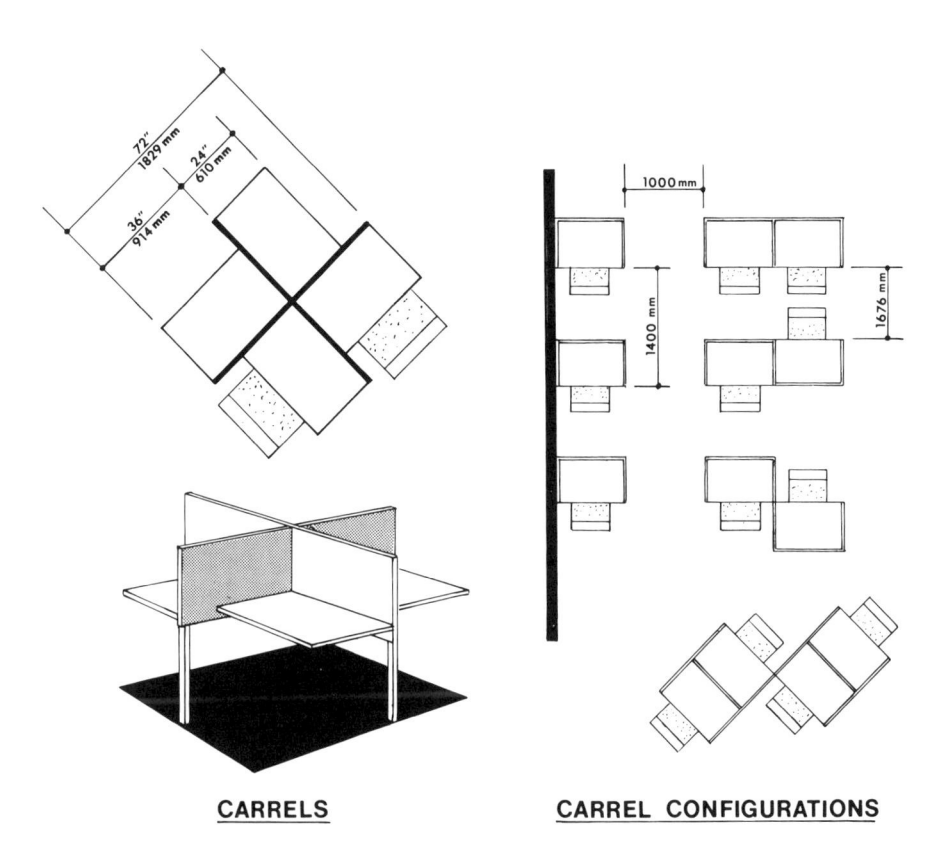

Reading Tables

Table sizes and spacing are each important to library planning. Multiple-seat tables should allow a minimum of 36 in (914 mm) width per seat. Spacing around and between reading tables is equivalent to spacing required for carrels as described above.

Present customary reading table sizes are:

	in	mm
RECTANGULAR		
Length (Four seats)	72	1829
Width	42	1069
	48	1219
Height	28	711
	28-1/2	724
	29	737

	in	mm
	SQUARE	
Four seats	42 × 42	1069 × 1069
	48 × 48	1219 × 1219
	ROUND	
Four seats (diameter)	54	1372
	60	1524

Book Trucks

Dimensions of book trucks (called "trolleys" in England) are relatively constant, whether they are constructed of wood or of metal, because dimensions relate to sizes of books; and moreover, maneuverability must be maintained in confined areas such as book stack aisles. Since these constants will remain, it is doubtful whether dimensions will change with metric conversion, other than rounding out awkward metric dimensions to create whole-metric sizes.

The following table gives customary dimensions and metric equivalents.

	in	mm
Length	28	711
	30	762
	31	787
	32	813
Depth	14	356
	15	381
	16	406
	17	432
	18	457
Height	36	914
	37-3/4	959
	38-3/4	984
	39-1/2	1003
	40	1016
	41	1041
	42	1067
	43	1092
Caster diameter	4	102
	5	127
	6	152

Index Tables

The same condition prevails with index tables, i.e., the basic dimensions will not change with conversion except to whole-metric sizes, because libraries will continue to hold existing book collections. Indexes and bibliographic material are standardized in dimension and it is doubtful if newly printed material will deviate considerably in size.

Customary dimensions of index tables are:

	in	mm
Length	60	1524
	90	2286
Width, single-faced	27	686
	30	762
Width, double-faced	54	1372
Height	38	965
	48	1219

Dictionary and Atlas Cases

Cases for dictionaries and atlases are designed to accommodate various sizes of volumes and thus can house adapted metric dimensions with little if any hard conversion of present measurements. Various customary dimensions are as follows:

	in	mm
Width	23-1/2	597
	24	610
	28-1/2	724
	30	762
Depth	14	356
	15-1/4	387
	24	610
	28-1/2	724
Height	43	1092
	44-3/4	1137

Reader Seating

Dimensions of chairs for optimum comfort when an individual is seated at a desk, table, or carrel vary little. Secretarial seating offers the luxury of swivel mechanisms, rolling casters, and adjustable seat

110 METRICS FOR ARCHITECTS, DESIGNERS, AND BUILDERS

and back controls, whereas library reading chairs (as well as dining chairs) are of fixed demension. Seat and back pitch, height, etc., are static dimensions.

	in	mm
ADULT		
Width of seat	16	406
	17-3/4	451
	18-1/2	470
	19	483
Height of seat	16-1/2	419
	17-1/2	445
	18-1/2	470
Depth of seat	15-1/2	394
	16-1/2	419
Width of back	12	305
	14	356
	16	406
Overall chair height	30	762
	31	787
	32	813
JUVENILE		
Width of seat	14-1/2	368
	15	381
Height of seat	13-1/2	343
	14	356
	15	381
Depth of seat	10-1/2	267
	11-1/2	292
	12	305
Width of back	9-1/2	241
	10	254
	11-1/2	292
Overall chair height	22-1/2	572
	23	584
	24	610

This poses a problem because of the variation in human sizes and proportions (see "Office Seating," Chapter 10). Chair makers have been mulling over this disparity for decades. Early academic and monasterial libraries presumed that mere benches encouraged good posture for reading and study. Today's competitive production demands comfort for library readers. Anthropometric studies have been made and conclusions reached about the optimum dimensions and proportions of library-reader seating. The table on p. 110 gives the metric equivalents of reader chair dimensions.

For adult-reader seating, clearance from the chair seat to the lower edge of table should be 7 to 9 in (178 to 229 mm). For juvenile readers, the clearance should be 6 to 7 in (152 to 178 mm).

Since the architectural planning of a library involves essentially open areas unrestricted by fixed partitions, much of the planning of interior spaces is oriented to the spacing and arrangement of furnishings, especially the bookstacks. Other than bookstacks as described above, furnishings can be adapted to metric building modules. This has been successfully done in a number of foreign libraries which were architecturally planned in metric modules but procured the furnishings in the United States, with English system measurements. Such adaptability is possible (again with the exception of bookstacks) because of the flexibility of large open areas.

13
Metrics in Residences

The style and character of residential furnishings are as varied as the often-fanciful design of the dwellings they occupy, which reflects the myriad tastes of the American public. Forms and sizes are innumerable; thus countless dimensions could be cataloged with metric equivalents. There are, however, certain common dimensional standards which predominate.

Two factors will become apparent in conversion to the metric system. One is that most home furnishings are not subject to architectural metric modules or standards. With some exceptions, such furniture is not dependent on building design adjustments to metric measure. Therefore hard conversion will essentially involve rounding dimensions to whole-metric sizes.

The other factor concerns seating, which is designed and proportioned to anthropometric measure and will remain so. If conversions are made in seating to whole-metric dimensions, such hard conversions will be no more than 1 or 2 mm.

Exceptions to this independence from metric building standards are:

1. Equipment or cabinetry that is built in and thus is a part of the architectural structure.
2. Loose cabinets, storage units, chests, etc., which can be more easily placed and fitted to spaces if they are dimensionally coordinated with accepted metric building modules.

Many home dwellers are avid collectors of antiques or authentic reproductions as well as some classic pieces of more recent vintage, such as the superb Barcelona chair. Such pieces, regardless of period, are precisely proportioned and dimensioned. To suggest rounding to whole-metric dimensions of such classic examples would bring cries of, "not a single millimeter."

With respect to the architecture of living spaces, some argue that conversion to SI metric standards is too regimental and restraining; that the home environment should rightfully be designed without a confining adherence to the 100-mm basic measure; that designers should be free to follow anthropometric dimensional relations. This, of course, is an excuse to avoid the challenge of creating within boundaries which may be considered restrictive. A livable house, apartment, or condominium can be designed just as well following accepted metrication principles—and with ingenuity and imagination.

Some opposers of the metric system contend that it is mathematically constituted and is therefore unhuman in concept. How so? What is the "human" difference between an object's being described as measuring one foot, 12 in, or 305 mm? It still fits the same shoe box. Perhaps this objection is made because the metric system is essentially rectilinear as well as modular and is not easily adapted to free-form structures (such as those of Antonio Gaudí). Nevertheless, in foreign countries *many* imaginative residences and other buildings have been created using the metric system of measure.

Following are customary living areas and their principal furnishings. Listed are various common dimensions as presently manufactured with equivalent metric conversions.

The Kitchen

Although kitchen cabinetry is often designed for the area and is built in on the job, many firms offer modular units prefabricated to specific sizes. The following table gives standard dimensions and their metric equivalents.

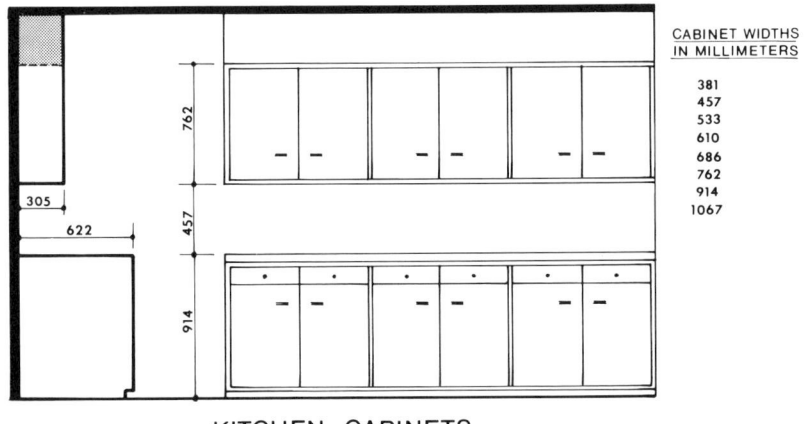

CABINET WIDTHS
IN MILLIMETERS

381
457
533
610
686
762
914
1067

KITCHEN CABINETS

114 METRICS FOR ARCHITECTS, DESIGNERS, AND BUILDERS

	in	mm
Counter height	36	914
Distance of counter to wall cabinets	18	457
Height of wall cabinets	30	762
Counter depth	24-1/2	622
Wall cabinet depth	12	305
Prefab cabinet widths	15	381
	18	457
	21	533
	24	610
	27	686
	30	762
	36	914
	42	1067
Typical sinks		
Length	24	610
	28	711
	30	762
	33	838
	42	1067
	46	1168
Width	16	406
	21	533
	22	559
	25	635
Refrigerators—freezers (Typical)*	24 × 27 × 59	610 × 686 × 1499
	32 × 29 × 66	813 × 737 × 1676
	35 × 31 × 66	889 × 787 × 1676
Typical micro-oven dimensions	24 × 19-3/4 × 15-3/4	610 × 502 × 400
	24 × 20-7/8 × 15-3/4	610 × 530 × 400
Washers—dryers (Typical)	24 × 20-1/2 × 41	610 × 521 × 1041
	24 × 25-1/2 × 43	610 × 635 × 1092
	29 × 27-1/2 × 45	737 × 699 × 1143

*12 ft³ = 0.340 m³ = 340 l (liters)
 17 ft³ = 0.481 m³ = 481 l
 19 ft³ = 0.538 m³ = 538 l
 22 ft³ = 0.623 m³ = 623 l

Dining

Dining is oftentimes a matter of an individual's mood and temperament. A person may eat at a breakfast bar, in an informal breakfast nook, or surrounded by the regalia of a formal dining room. The furnishings and their dimensions are designed and dimensioned accordingly. In many European nations dining-table heights are lower than is the American standard of 29 in, or 737 mm. Accepted dining table heights in other nations are often as low as 660 mm (26 in).

Extensive tests have determined that 737 mm is too high for comfortable dining or working. From anthropometric studies, the following table heights have been recommended, as long ago as the Hooton research studies done at Harvard in 1945:

mm	in
660	26
700	27.5
710	28
720	28.3

The latter dimension is the accepted maximum height for comfortable dining (or working).

It may be that the retention by American manufacturers of the greater dimension in table heights stems from the days when table construction required deep aprons which restricted knee room. At one time, table heights were 30 in (762 mm), and even 31 in (787 mm). With the conversion to the metric system, it is hoped that table makers will adjust to the metric dimensions indicated above. Present standards in most foreign countries for dining table heights are 710 and 720 mm. The "continental height" of 660 mm is a compromise between the kaffee klatsch and dining.

Service bar heights—breakfast or otherwise—are:

42 in, or 1069 mm, which may be rounded to 1050 mm
45 in, or 1143 mm, which may be rounded to 1150 mm

A normal place setting occupies approximately 24 × 15 in (610 × 381 mm). Table seating allowance for dining should be no less than 24 in or 610 mm per chair (armless), and a minimum of 27 in (686 mm) per armchair. For seating at ends of a table allow an additional 460 mm (18 in) at each end.

DINING

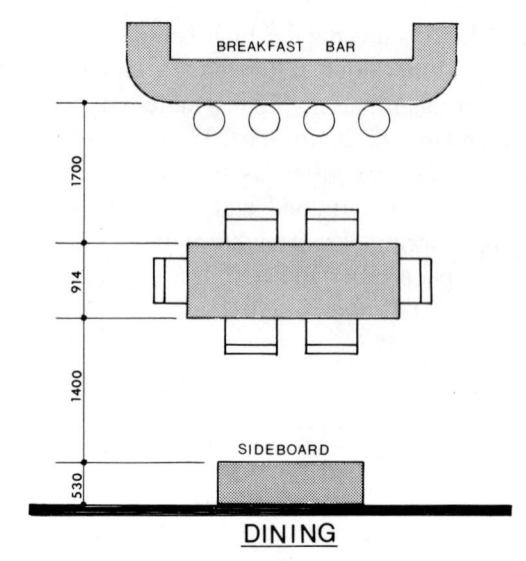

The following gives present customary dining-table sizes along with metric equivalents. Note that a table leaf or extension will add 12 in (305 mm) or 18 in (457 mm) to a table length.

in	mm	NUMBER OF SEATS
RECTANGULAR		
54 × 30	1372 × 762	4
60 × 30	1524 × 762	4
72 × 36	1829 × 914	6
78 × 36	1981 × 914	6
84 × 36 (extended)	2134 × 914	6–7
90 × 36 (extended)	2286 × 914	8
96 × 36 (extended)	2438 × 914	8–10
SQUARE		
30 × 30	762 × 762	2–4
32 × 32	813 × 813	4
36 × 36	914 × 914	4
42 × 42	1067 × 1067	4
48 × 48	1219 × 1219	4
ROUND		
30 (diameter)	762 (diameter)	2
36	914	2–4
42	1067	4
48	1219	4
60	1524	4–6

Sideboards

A sideboard or English "buffet" may provide serving space as well as storage. Following are common customary length and width dimensions:

in	mm
32 × 18	813 × 457
36 × 18	914 × 457
42 × 21	1067 × 533
48 × 21	1219 × 533
66 × 21	1676 × 533
72 × 21	1829 × 533
84 × 21	2134 × 533
90 × 21	2286 × 533

The height may be any reachable dimension but normally not over 72 in (1829 mm) or 84 in (2134 mm). If a sideboard is to be used for service the height should be from 36 in (914 mm) to 42 in (1067 mm).

Seating

Requirements for comfortable seating at dining tables are as described in Chapter 10, "Metrics and the Business Office," and Chapter 12, "Metrics and Libraries." It should be noted, however, that the distance from the seat height to the table top should not exceed 230 mm (9 in). Thus, if dining-table heights are converted to the preferred metric dimensions indicated in the preceding paragraphs, dining-chair-seat heights should be adapted accordingly:

TABLE HEIGHT, mm	SEAT HEIGHT, mm
660	430
700	470
710	480
720	490

Armrests should clear the underside of the table top or be shortened to permit sliding the chair up to the table.

Living

Casual living is today's theme, which implies informality and lighter-scaled furniture. Present customary sizes of furniture have followed this trend. There is greater emphasis on flexibility, which means more use of modular units, in seating as well as in storage. Also this means that the average living or family room is less cluttered and contains fewer pieces of furniture. Conversion to metric modular coordination in building design hence presents less of a problem in planning living areas, even with furniture as presently dimensioned.

Following are standard items of living furniture, with customary dimensions and metric equivalents.

	COFFEE TABLES	
	in	mm
Heights	14	356
	15	381
	16	406
	17	432
	18	457
Rectangular	24 × 18	610 × 457
	30 × 20	762 × 508
	36 × 20	914 × 508
	40 × 20	1016 × 508
	60 × 20	1524 × 508
	60 × 24	1524 × 610
	60 × 30	1524 × 762
	72 × 24	1829 × 610
	72 × 30	1829 × 762
	78 × 24	1981 × 610
	78 × 30	1981 × 762
Square	24 × 24	610 × 610
	30 × 30	762 × 762
	32 × 32	813 × 813
	36 × 36	914 × 914
	42 × 42	1067 × 1067
	48 × 48	1219 × 1219
Round (diameter)	24	610
	30	762
	36	914
	42	1067
	48	1219

COFFEE TABLES—Continued

	in	mm
Oval	44 × 34	1118 × 864
	54 × 36	1372 × 914
	60 × 30	1524 × 762
	70 × 36	1778 × 914
	72 × 34	1829 × 864
	78 × 48	1981 × 1219

Occasional Tables

Dimensions are approximately the same as those for smaller coffee tables except heights are between 20 in (508 mm) and 22 in (559 mm).

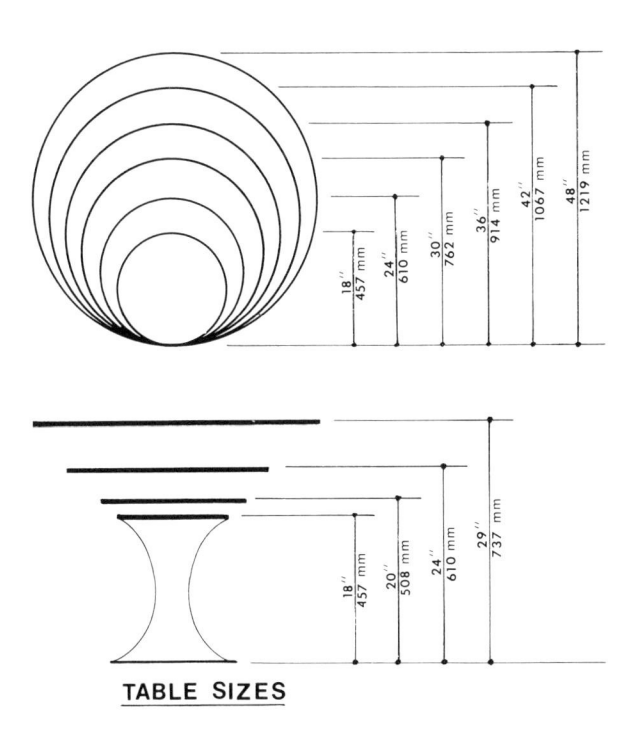

TABLE SIZES

Lounge Chairs

It is hard to say if today's casual posturing in lounge chairs is the cause or effect of our modern, informal, and sometimes sprawling chair forms. Traditional lounge chairs such as the wing chair, the fireside, the barrel, the fanback, and the tub chair encouraged proper sitting with higher seats and gently sloped but more erect backs.

The shapes and contours of today's lounge chairs are infinite, but their proportions commonly indicate lower seat heights and often deeply inclined backs. The day of the prim teatime fireside chat is past. Stiff corsets and buckram coats have no bearing on the ductile forms. Now the theme is relaxation, with or without TV sedatives. The shape of today's lounge chairs is tacit proof of this.

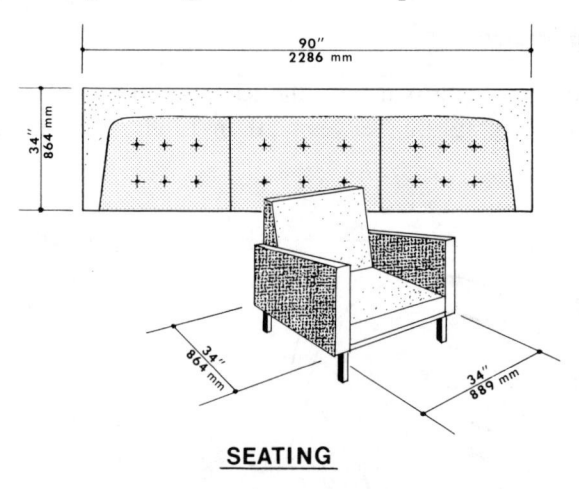

SEATING

The following table gives common dimensions of lounge chairs, including customary seat heights.

	in	mm
Width	27	686
	28	711
	30	813
	32	864
	34	914
	36	1016
	40	
Depth	27-1/2	699
	30	762
	33	838
	34	864
	36	914
	40	1016

	in	mm
Height	25	635
	26-1/2	673
	28	711
	33	838
	39-1/2	1003
Seat height	13-3/4	349
	15-1/2	394
	16	406
	16-1/2	419

Sofas and Modular Seating

Sofas can generally be classified as tuxedo, Lawson, roll-arm, and, of course, armless, with loose cushions or tight seat. Seat heights vary according to the depth and resilience of the cushioning, with tight seat units indicating a lower height of the seat than sofas with soft, deep cushioning. Seat heights are in the following range:

in	mm
14	356
15	381
16	406
17	432

Sofa lengths vary because of differences in design and construction, as well as the presence or absence of arms. A common sofa width is 90 in (2286 mm) with minor variances, and a few manufacturers offer oversized sofa units up to 108 in (2743 mm). Because lengths are simple for manufacturers to adjust, such dimensions can easily adapt to hard conversion, i.e., the measurements above can be translated to 2200 or 2300 mm, and 2700 or 2750 mm. Present sofa dimensions are as given in the following table (p. 122).

Modular seating, as the name implies, is a composition of assembled units. Such seating is alleged by some to be the interior designer's answer to flexibility in planning. With inside and outside wedge-shaped seating modules as well as straight units, innumerable contours and configurations are possible. Any number of seating units can be employed in the assembly.

		in	mm
Width	(Settee)	55	1397
	(Settee)	59	1499
	(Settee)	61	1549
	(Settee)	66	1676
		72-1/2	1842
		80	2032
		84	2134
		87	2210
		89	2261
		90	2286
		94	2388
		96	2438
Depth		30	762
		32	813
		33	838
		34	864
		36	914
		37	940
Height		27	686
		28	711
		30	762
		32	813
		34	864

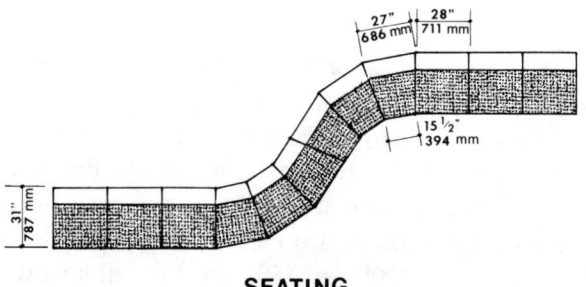

SEATING

The concept is especially suitable for large areas such as public waiting rooms and lounges, but it is equally adaptable to residential living areas. One advantage of modular seating, according to manufacturers, is that the units, joined by connector devices, can be rearranged and added onto.

The customary angle of wedge-shaped seating modules is either 22-1/2° or 30°, with the back at either the inside or the outside of the wedge.

Standard seat depths are customarily 31 in (787 mm) or 34 in (864 mm). Following are customary widths:

	in	mm
Straight module	28	711
	30	762
Wedge (22-1/2°)		
Outer	27	686
Inner	15-1/2	394
Wedge (30°)		
Outer	27	686
Inner	12	305

Since modular seating assemblies may assume various formations and curvatures, there is little point in urging hard conversion to building modular standards. For simple convenience to interior designers working in metrics, however, dimensions may be converted to whole-metric measurements, such as straight modules at 700 mm and 800 mm, etc.

Television

Sizes of TV screens and consoles vary widely among manufacturers. Screen dimensions, measured diagonally, range from a minute 2 in (51 mm) to a huge projection screen which measures 50 in (1270 mm). Following are some present customary screen sizes:

in	mm
5	127
9	229
12	305
19	483
21	533
25	635

Overall dimensions include the following:

	in	mm
Width		
Portable	8	203
	12	305
	16	406
	22	559
Table	25	635
	27	686
Console	32	813
	40	1016
	48	1219
	68	1727
With stereophonic speakers	72	1829
Depth	10	254
	15	381
	19	483
	22	559

Stereos

Most audio equipment today is marketed in components which offer a myraid of sizes and assembly configurations, and therefore dimensions. Stereo ensembles housed in console cabinets are available however. Following are some common dimensions:

	in	mm
Width	38	965
	48	1219
	54	1372
	60	1524
	72	1829
Depth	16-1/2	419
	17	432
	19-1/2	495

Standard phonograph record albums measure 13-1/4 × 13-1/4 in (337 × 337 mm), or 13-1/2 × 13-1/2 in (343 × 343 mm).

The customary width and depth of a turntable unit is 13-1/2 × 12 in (343 × 305 mm).

Speakers are often placed separately in order to achieve stereo sound fidelity. Common dimensions of individual speakers are:

	in	mm
Width	8-1/2	216
	10	254
	11-1/2	292
	12	305
Depth	7-1/2	191
	8-1/2	216
	9-7/8	251

Pianos

Having a baby grand in the parlor or living room is a thing of the past. Since the innovation of the jazz age and later rock rhythm, smaller spinet or studio pianos have become accpeted furniture components in living areas, particularly family or recreation rooms. Dimensions of smaller upright pianos are similar to the popular do-it-yourself electric organs.

Hard conversion to metric measure may alter the overall dimensions of keyboard instruments, even though the mechanical dimensions will undoubtedly remain. Following are present customary dimensions:

	in	mm
Baby Grand		
Width	62	1575
Depth	58	1473
Spinet—Studio		
Width	58	1473
Depth	24-1/2	622
	24-3/4	629

Sleeping

The focal point of the bedroom has traditionally been the bed, often four-postered and canopied. Dimensions were therefore formidable.

126 METRICS FOR ARCHITECTS, DESIGNERS, AND BUILDERS

Today the bed is less monumental, the emphasis being more on comfort. Present mattress sizes determine overall bed sizes. Moreover, the bedroom sometimes doubles as a den/library/lounge area.

Conversion of mattress sizes to whole-metric dimensions will not be affected by SI metric building standards, except when a bed must fit into an alcove. The following are present standard mattress sizes:

	in	mm
Twin	75 × 39	1905 × 991
	80 × 39	2032 × 991
Three-quarter	75 × 48	1905 × 1219
Double	75 × 54	1905 × 1372
	80 × 54	2032 × 1372
Queen size	80 × 60	2032 × 1524
King size	80 × 72	2032 × 1829
	80 × 76	2032 × 1930
	84 × 72	2134 × 1829
	84 × 76	2134 × 1930

The overall dimension of a bed will of course be increased by the thickness of head and footboards.

Chest or dresser storage sometimes is comprised of multiple units, often including a vanity table. Because such units must often fit into an architectural dimension, singly or in multiples, the present customary dimensions will undoubtedly be converted to conformity with metric building standards, rounding to multiples of the 100-mm basic measure. Present customary dimensions are as follows:

	in	mm
Width	24	610
	30	762
	32	813
	36	914
	42	1067
	48	1219
	60	1524
	72	1829
	80	2032

	in	mm
Depth	17-1/2	445
	18	457
	20	508
	21	533
Height	29-1/2	749
	32	813
	34	864

Night stands or tables are generally small and incidental, often incorporated as part of a wide headboard. Dimensions are therefore nominal:

	in	mm
Width	18	457
	20	508
	22	559
	24	610
Depth	15	381
	17	432
	18	457
Height	22	559
	23-1/2	597
	25	635

Juvenile bedding has its own standard of sizes which varies according to age brackets. Following are standard children's mattress dimensions:

in	mm
36 × 18	914 × 457
38-1/2 × 22-1/4	978 × 565
46 × 23	1168 × 584
50-1/2 × 25-1/4	1283 × 641
51 × 27	1295 × 686
56-1/2 × 31-1/4	1435 × 794

Some Common Furniture Measurements

in		mm	ROUNDED TO	mm		in
12	=	305		300	=	11.8
15	=	381		400	=	15.7
18	=	457		450	=	17.7
20	=	508		500	=	19.7
24	=	610		600	=	23.6
30	=	762		750	=	29.5
36	=	914		900	=	35.3
42	=	1067		1050	=	41.3
48	=	1219		1200	=	47.2
60	=	1524		1500	=	59.1
72	=	1829		1850	=	72.8
84	=	2134		2150	=	84.6
90	=	2286		2300	=	90.6

14
Auditoriums and Assembly

Assembly areas with fixed seating, such as most theaters and auditoriums, must conform to certain safety codes that limit the number of seats between aisles. In the United States, the allowable maximum is 13 to 15 seats between aisles. In other countries such codes do not apply; and "continental" seating plans often employ as many as 50 to 100 seats between aisles.

In the seating plan, space must be provided for wheelchair handicapped at a ratio of 1 percent; or in small assembly areas, a minimum of one available space (see Chapter 15 "Metrics and the Handicapped").

Aisle widths should be no less than 1000 mm (39.4 in), and should increase in width from front to rear at a ratio of 1:50. Where handicapped codes apply, the minimum width should be no less than 1067 mm (42 in).

Recommended minimum spacing between seating rows, measured back to back, is 32 in (813 mm), with preferred spacing 34 to 36 in (864 to 914 mm).

Customary widths of fixed seating units are:

in	mm
19	483
20	508
21	533
22	559

Some building codes permit no more than 22 seating rows without intermediate cross aisles, which should be a minimum of 1067 mm in width and preferably no less than 1524 mm (60 in) to allow pivoting of wheelchairs.

Many smaller assembly rooms are multipurpose, to be used occasionally for displays and exhibits or seminars and conferences with

130 METRICS FOR ARCHITECTS, DESIGNERS, AND BUILDERS

folding tables. Stage platforms are oftentimes portable and removable. Floors must be level, not sloped.

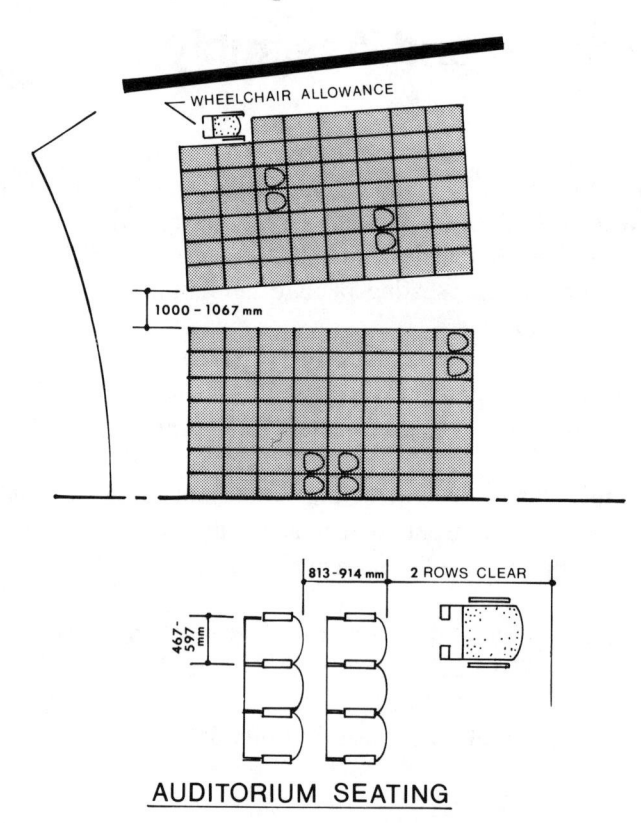

AUDITORIUM SEATING

Seating for such rooms must be designed for easy removal and compact storage when not in use. Stacking chairs that can be aligned by ganging devices are most commonly used. Wheeled dollies permit stacking as many as 45 chairs. The chairs can then be rolled into storage rooms. Dimensions of customary stacking chairs are as follows:

	in	mm
Width	18-3/8	467
	19-1/2	495
	21-1/2	546
	23-1/2	597

sons. Pivoting and maneuvering a wheelchair requires certain dimensional clearances and spatial allowances, which are spelled out in all handicapped codes and regulations. These dimensional standards must, of course, be converted to metric equivalents.

The standard dimensions of wheelchairs and their maneuverability will not change with metric conversion, because the design, proportions, and operating are anthropometric, not architectural.

Following are customary dimensions of adult wheelchairs with their metric equivalents:

	in	mm
Width	27	686
Length	42	1067
Height	34	864
Armrest height	29	737
Seat height	19-1/2	495

Turning Radii

A wheelchair may be pivoted from either the central pivot point or the axis of either wheel.

- From the central pivot point, the turning radius is 31.5 in, or 800 mm. The track radius of the small caster wheels is 19 in, or 483 mm.
- From the axis of either wheel, the pivoting radius is 36 in, or 914 mm. The track radius of the small caster wheels is 18 in, or 457 mm.
- The minimum radius allowance for a complete 360° turnaround is 54 in, or 1372 mm.

Wheelchair Reach

- Lateral reach from the center of the wheelchair averages 32 in, or 813 mm.
- Vertical reach is an average of 62 in, or 1575 mm. However, an angled upward reach to bookshelves, for example, will reduce this height to approximately 59 in, or 1499 mm.

136 METRICS FOR ARCHITECTS, DESIGNERS, AND BUILDERS

- Downward reach will average approximately 13 in, or 330 mm, from the floor.
- Forward reach, such as access to writing or work surfaces, will average from 24 in, or 610 mm, to 30 in, or 762 mm, depending on the activity.
- Normal reaching height for telephones, switches, etc., is 40 in, or 1016 mm, from the floor.

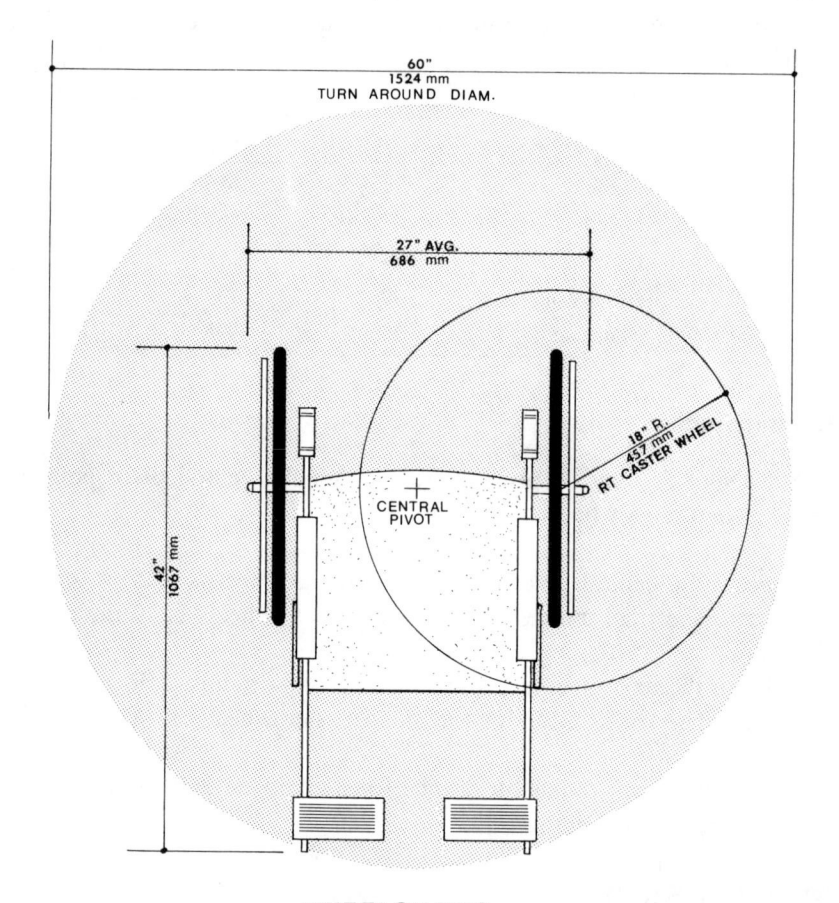

WHEELCHAIRS

Space Allowances

Certain minimum clearances and area dimensions are necessary in order to allow wheelchair users to move and maneuver freely. In pub-

METRICS AND THE HANDICAPPED 137

lic areas these minimum requirements are vital to a barrier-free environment. All handicapped codes require that these conditions be met. The most critical, with metric equivalents, are:

- A minimum width of 48 in, or 1219 mm, is required for walkways, corridors, and passageways. Preferred is 60 in, or 1524 mm.
- The preferred width of 1524 mm will permit a wheelchair turnaround. In addition, this width will allow two wheelchairs to pass each other.
- Door-width clearances should be no less than 32 in, or 813 mm. Preferred is 36 in, or 914 mm. This dimension applies also to elevator doors.
- Between two sets of doors, a minimum of 78 in, or 1981 mm, should be allowed.
- Minimum size of public elevators should be 66 in wide by 61 in deep, or 1676 mm wide by 1549 mm deep. Residential elevators should be no less than 40 in wide by 52 in deep, or 1016 mm wide by 1321 mm deep.
- Parking spaces for the handicapped (which are normally designated as such) should be no less than 150 in, or 3810 mm wide to permit wheelchair accessibility.
- Bedrooms should allow wheelchair turnaround space. In addition, the space between twin beds should be 40 in, or 1016 mm; and that between bedside and wall, no less than 54 in, or 1372 mm. The clearance from the foot of the bed should be a minimum of 48 in, or 1219 mm.
- Bathrooms and restrooms should allow turnaround space of 60 in, or 1524 mm, square. Toilet stalls should provide a clear width of 42 in and depth of 72 in, or 1067 mm by 1829 mm.

Other Wheelchair-related dimensions

- Work-counter usable depth is approximately 24 in, or 610 mm. In order to clear the standard wheelchair armrest height of 737 mm, the counter height should be 31 in, or 787 mm.
- The minimum width per occupant at work counters (clear) should be 30 in, or 762 mm.
- The customary height of railings for all handicapped is 32 in, or 813 mm.

138 METRICS FOR ARCHITECTS, DESIGNERS, AND BUILDERS

- Maximum slope of all ramps should be a ratio of 1:12, with a level rest platform at intervals of no more than 360 in, or 9144 mm.

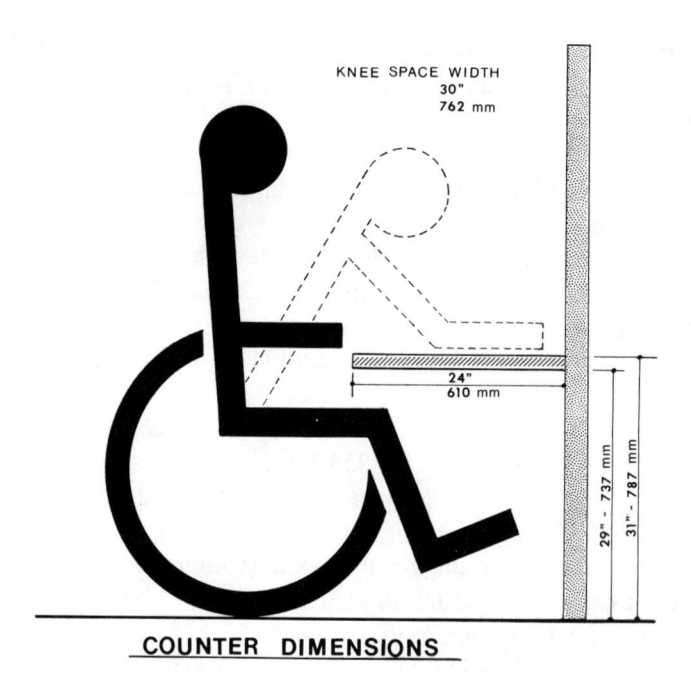

KNEE SPACE WIDTH
30"
762 mm

24"
610 mm

29" - 737 mm
31" - 787 mm

COUNTER DIMENSIONS

There are of course many additional requirements to make public facilities and buildings accessible to the physically handicapped which are not noted here. However, the purpose here has been to cover only *dimensional* requirements, giving customary measurements with their metric equivalents.

Selected Bibliography

American National Metric Council, "Metric Conversion in the U.S. Construction Industries," *ANMC Metric Reporter*, vol. 8, no. 7, 1980. 5410 Grosvenor Lane, Bethesda, Md. 20014 (subscription $50/yr).

American National Standards Institute, *Standard Practice for the Use of Metric (SI) Units in Building Design and Construction,* ANSI/ASTM E 621-78, 1978.

American National Standards Institute, *Specifications for Making Buildings and Facilities Accessible to and Usable by Physically Handicapped People,* 1980.

Braybrooks, Susan, (ed.), *AIA Metric Building and Construction Guide*, Wiley, New York, 1980.

Diffrient, Niels; Tilley, Alvin R.; and Bardagjy, Joan C., *Humanscale 1/2/3*, M.I.T., Cambridge, Mass., 1974.

Fairweather, Leslie; and Sliwa, Jan A., *VNR METRIC HANDBOOK*, Van Nostrand, Princeton, N.J., 1969, rev. 1977.

Harkness, Sarah P.; and Groom, James N. Jr., *Building without Barriers for the Disabled*, Whitney Library of Design, New York, 1976.

Mace, Ronald L.; and Laslett, Betsey, "An Illustrated Handbook of the Handicapped Section of the North Carolina State Building Code," North Carolina Building Code Council and North Carolina Department of Insurance, 1979.

Milton, Hans J. (ed.), *Recommended Practice for the Use of Metric (SI) Units in Building Design and Construction*, National Bureau of Standards, SD Stock No. 003-003-01761-2, 1977.

National Bureau of Standards, *Units and Systems of Weights and Meaures, Their Origin, Development and Present Status,* LC 1035, 1960, amended 1976.

Packard, Robert T. (ed.), *Architectural Graphic Standards*, Wiley, New York, 1981.

Pedde, Lawrence D.; Foote, Warren E.; Scott, LeRoy F.; King, Danny L.; and McGalliard, Dave L., *Metric Manual*, U.S. Department of the Interior Bureau of Reclamation, SD Stock No. 024-003-00129-5, 1978.

Rubin, Arthur I.; and Elder, Jacqueline, *Building for People*, Environmental Design Research Division Center for Building Technology, National Engineering Laboratory, National Bureau of Standards, 1980.

Wilson, Forrest, *Structure: The Essence of Architecture*, Van Nostrand, Princeton, N.J., 1971.

Index

Aisles, widths and clearances, 99, 100, 101, 102, 106, 129, 132
American National Metric Council, ANMC, 6, 11, 12, 60
American National Standards Assoc., ANSI, 58, 133
Angles, steel, *see* Metal
Anthropometric (anthropometry), vi, 10, 25, 52, 54, 56, 57, 82, 87, 94, 111, 112, 113, 115, 135
Architects, v, vi, 5, 7, 21, 22, 42, 48, 49, 58, 59, 60, 78, 106
Architectural, dimensions, 21, 47, 73, 126; planning, 13, 14, 111, 131; practice, 65; proportion, 51; scale, 37, 38, 39; standards, 18, 79
Architectural Barriers Act of 1968, Pub. Law 90-480, 133. *Also see* Handicapped
Area, 17, 26, 31, 32, 33, 34, 35, 43, 44, 60, 66, 113, 134, 136
Assembly, *see* Auditoriums
Atlas stands, *see* Dictionary
Auditoriums, 129, 130, 131

Barcelona chair, 56, 112
Barcelona, Spain, 3
Bars, breakfast and restaurant, 95, 96, 98, 114, 116
Beams, *see* Metal, *also see* Wood
Beds, sizes of, 25, 125, 126, 127, 137
Blocks, concrete, *see* Masonry
Bookstacks, library, *see* Shelving
Book trucks, library, 108
Brick, *see* Masonry
British Brick Development Assoc., BDA, 61
British Standards Institution, BSI, 11
Builders, 9, 13, 41, 59
Building, design, 7, 14, 18, 24, 25, 42, 44, 52, 53, 58, 60, 76, 89, 91, 99, 112, 118, 134; construction, 9, 77, 134; industry, 11, 13; materials, 60, 61; modules, 67, 79, 87, 91, 94, 100, 112, 132; products, 13, 22, 24

Cabinets, cabinetry, 18, 40, 59, 112, 113, 124
Card catalogs, library, 105
Carpet, 17, 23, 59, 76, 77; gauge of, 76
Carrels, library, 106, 107
Ceiling panels, *see* Panels
Centimeter, 15, 17, 18, 21, 37, 44
Chairs, *see* Seating
Channels, steel, *see* Metal
Column size, 102
Computer Output Microfilm, COM, 105
Construction, 7, 17
Conversion, v, 7, 9, 11, 13, 14, 18, 20, 21, 22, 25, 47, 112; hard, 6, 7, 10, 11, 13, 22, 23, 51, 53, 57, 58, 60, 65, 72, 73, 79, 82, 85, 91, 94, 99, 109, 122, 123, 125, 132; metric, v, 13; modular, 23, 24; soft, 22, 23; voluntary, v, 4, 6, 9, 13
Coordination, design, 7, 25; dimensional, 7, 14, 23, 25; modular, 14, 23, 25
Counters, 55, 57, 95, 114, 137
Credenzas, 84
Cubit, 1

da Vinci, Leonardo, 51
Decimal, 22, 43; fractions, 18, 21, 23; place, 18, 41, 44; system, 7, 15, 37
Decking, 83
Department of Housing and Urban Development, HUD, 60
Designers, v, vi, 5, 7, 8, 14, 21, 22, 38, 42, 46, 47, 48, 49, 55, 56, 78, 106, 112, 113, 122, 123
Desks, 10, 22, 47, 54, 57, 79, 81, 82, 84, 87, 92; library, 105, 106
Dictionary-atlas stands, library, 109
Dimensional, 8, 58; architectural, 21; coordination, 11, 58; conversions, 23; standards, 8, 58, 73, 84, 113
Dimensions, 112, 113, 118; customary, 24, 60, 72, 79, 92, 99, 105, 106, 108, 109, 118, 126, 131, 135; metric, 23, 47

141

Dining, 95, 96, 114, 115, 116, 117
Disabled American Veterans, DAV, 134
Divine Proportions, see Golden Mean
Doors, v, 25, 47, 58, 59, 72, 137
Duodecimal system, 1, 22

English system of measure, 1, 2, 4, 5, 6, 14, 17, 18, 20, 21, 22, 23, 24, 25, 37, 39, 46, 53, 56, 58, 60, 62, 77, 111
Ergonomics, consideration of, 55
Euclid, 50
European Common Market, 5

Fibonacci, 51, 52
Filing cases, 10, 18, 23, 47, 79, 84, 92
Flooring, 66
Fractions, 1, 2, 15, 17, 37, 39, 43, 76; decimal, 21, 23, 65, 76
French Academy of Sciences, 3
French Revolution, 3
Furniture, v, vi, 5, 10, 14, 18, 22, 24, 40, 46, 53, 54, 55, 56, 58, 59, 79, 111, 114, 118, 125

Gallon, British imperial, 2
Gaudi, Antonio, 113
Gauges, metal, 70, 71, 76; carpet, 76
General Conference on Weights and Measures, 4
Golden Mean, proportions, rectangle, Rule, spiral, 51

Handicapped, 100, 101, 102, 104, 106, 129, 131, 133, 134, 137, 138
Hectare, 16, 17
Hooton, Dr. Earnest Albert, *A Survey in Seating,* Harvard U., 1945, 82, 115
HPL, high pressure laminate, see Panels
Human dimensions, 52, 55, 87; form, 56; measure, 53, 54, 57; proportions, 51, 53
Humanscale, 82

Inches-feet, v, 2, 5, 8, 17, 24, 25, 46, 47, 48, 76, 82, 93, 94, 99, 113
Index tables, library, 109
Interior designers, see Designers
International Bureau of Weights and Measures, 3
International Metric Convention, 3
International Organization for Standardization,
ISO, vi, 8, 10, 25, 41, 59, 61, 62, 80, 84, 105
International Year of the Handicapped, 134
Inventory control, 7

Joists, see Wood

King Henry I, 2
Kitchen, 113, 114

Le Corbusier, 51
Library, 80, 99, 100, 101, 104, 105, 106, 107, 109, 110, 111, 126
Louis XVI, King, 3
Lumber, 2, 58, 59, 62

Man, measure of, 50, 51, 52, 53
Masonry, 58, 59, 61, 62
Materials, building, 7, 61; fabricated, 7, 55, 58, 62, 106; structural, 67
Measure, area, 17, 43, 44; cubic, 25; customary, 92, 138; dry, 2; human, 55; linear, 15, 16; metric, 21, 22, 23, 24, 41, 47, 48, 49, 55, 56, 59, 60, 65, 71, 73, 77, 87, 89, 91, 92, 99, 125; standard, 46; system of, 2, 3, 4, 5, 6, 7, 8, 9, 10, 11, 14, 21, 23, 24, 50, 52, 58, 60, 111, 113; units of, 1, 2, 17, 20, 40, 44, 48; volume of, 2, 44
M-day, 11
Metal, 58, 67, 68, 69
Metcalf, Dr. Keyes D., 99
Meter, International Prototype, 3
Meter, Treaty of the, 3
Metric conversion, v, 6, 7, 8, 10, 13, 17, 21, 22, 23, 24, 25, 73, 78, 99, 101, 104, 108, 109, 135; dimension, v, 4, 5, 7, 9, 10, 22, 25, 42, 47, 55, 58, 59, 60, 61, 67, 89, 108, 109, 116, 117; equivalents, 20, 60, 62, 66, 79, 87, 95, 104, 108, 112, 113, 116, 118; measure, 3, 9, 24, 41, 46, 47, 48, 49, 55, 56, 60, 65, 73, 77, 84, 89, 91, 99, 125; modules, 8, 10, 24, 59, 62, 66, 111, 118, 132; scale, 37, 38, 46; sizes, 85; standards, vi, 40, 60, 77, 112, 126; system, v, vi, 3, 4, 9, 11, 17, 21, 24, 40, 47, 52, 56, 58, 59, 77, 78, 80, 99, 112, 115; units, 14, 40, 42, 48, 79, see also, Tables, conversion

Metric Conversion Act of 1975, v, 4, 6, 60
Metric, whole, 59, 60, 79, 87, 89, 96, 99, 108, 109, 112, 123, 126
Metricated, 6, 11
Micro-ovens, 150
Mile, 25; German geographical, 2; nautical, 2; statute, 2; survey, 2
Modular, 24, 25, 60, 66, 67, 77, 79, 84, 89, 91, 94, 99, 100, 102, 111, 112, 113, 118, 122, 123, 132; coordination, 7, 23, 77; standardization, 7, 77
Modular Building Standards Assoc., MBSA, 58

National Bureau of Standards, 4, 58
National Institute of Building Sciences, NIBS, 6
Nervi, architect, 77

Orientation, v, 5, 7, 8, 9, 11, 46, 47, 48

Panels, 23, 56, 58, 59, 62, 65, 67, 94; drywall, 10, 59, 62, 64; HPL, 65, 66; plywood, 10, 22, 56, 59, 62, 64; wallboard, 22, 23, 62
Paper, 80, 81, 82, 84, 85, 105
Phonograph, *see* Stereo
Pianos, 125
Planners, *see* Designers
Plywood, *see* Panels
Prefix, 15, 16, 17
Pythagoras, 50

Refrigerators, 149
Remodeling, v, 8, 133, 134
Rounding, 10, 22, 23, 25, 47, 53, 59, 61, 76, 96, 100, 105, 112, 126, 132
Rules and recommendations, 42

Scale, 37, 38, 39, 40, 46
Schedule, time, 11, 12, 13
Seating, 10, 11, 47, 53, 56, 57, 94, 131
   Auditorium, 129, 131
   Conference, 87, 89
   Dining, 110, 116, 117
   Executive, 87
   Folding—stacking, 130, 131, 132
   Library, 106, 107, 109, 110, 111
   Modular, 121, 122, 123
   Office, 87, 88, 110

Residential, 112, 119, 120, 121, 122
Restaurant, 95, 96, 97, 98
Secretarial, 87, 109
Wheelchair, 100, 101, 106, 129, 131, 134, 135, 136, 137
Senior citizens, 134
Sèvres, France, 3
Shelving, library, 58, 99, 100, 101, 102, 103, 104, 111, 135
SI, International System of Units, v, vi, 4, 5, 6, 9, 10, 14, 16, 18, 20, 25, 37, 40, 42, 44, 46, 55, 59, 79, 112, 126
Sideboards, 117
Sinks, kitchen, 114
Sleeping, *see* Beds
Spacing, 59, 79, 96, 99, 100, 102, 104, 106, 107, 129, 131, 132
Specifications, 24, 42, 44, 77
Standardization, modular, 7; *also see* Modular
Standards, dimensional, 8, 58, 112; modular, 7, 8, 14, 94; *also see* Modular
Steel, *see* Metal
Stereos, 124
Structural, lumber, 58, 62; pipe, 68; steel, 58, 67; tubing, 68
Stockpiling, v, 8, 9, 60
Symbols, metric, 42

Tables, 47, 57, 79; conference, 89; dining, 96, 115, 116; folding, 130; index, 109; library, 107; occasional, 118, 119, 127; restaurant, 95, 96, 97; work, 82, 137
Tables, conversion,
   Acres to hectares, 35
   Anthropometric dimension, 54
   Areas, 32
   Bed sizes, 126, 127
   Book trucks, library, 108
   Brick sizes, 61
   Card catalogs, library, 105
   Carpet, 77
   Ceiling panels, 67
   Chests, dressers, 126, 127
   Concrete block sizes, 83, 106
   Decking, 63
   Dictionary—atlas cases, sizes, 109
   Door sizes, 72, 73
   Ft to m, 29
   Ft to mm, 28

Tables, conversion (*continued*)
   Ft^2 to m^2, 33, 34
   Ft^3 to m^3, 36
   Filing case sizes, 85, 86
   Flat steel, 69, 70
   Flexible wall coverings, 71, 72
   Flooring, resilient, 66
   Furniture sizes, common, 128
   Gauges, metal, 70, 71
   HUD living sizes, 61
   In to mm, 27
   In^2 to mm^2, 31
   Kitchen cabinet sizes, 114
   Lumber sizes, 63
   Mi to km, 30
   Movable wall systems, 92, 93
   Panel sizes, 64, 66, 92
   Paper sizes, 80, 81
   Piano sizes, 125
   Pipe and tubing dimensions, 68, 69
   Seating dimensions, 88, 96, 110, 120, 121, 122, 130, 131
   Sideboard sizes, 117
   Stereo sizes, 124
   Structural steel, 67
   Table sizes, 90, 95, 107, 108, 109, 116, 118, 119, 127
   Tables, index, library, 109
   TV sizes, 124
   Wheelchair sizes, 135
   Window dimensions, 73, 76
TV (television), 120, 123, 124
Tubing, *see* Metal

U.S. Metric Board, 4

Van der Rohe, Mies, 56
Visualization, 46, 47, 48, 49
Vitruvius, 77
Volume, spatial, 2, 36, 44
Voluntary, v, 4, 9

Wallboard, *see* Panels
Wall coverings, 59, 71, 72
Wall systems, movable, 91, 92, 93, 94
Washers—dryers, 114
Wheelchairs, *see* Seating
Windows, v, 47, 58, 59, 73, 74, 75, 76
Wood, framing, 62; joists, 62; studs, 62
Work surfaces, 92, 93, 94, 96, 136, 137

Yard (measure), 1, 2, 25, 72; square, 23, 72, 77